AMERICAN BOOK COMPANY'S

MASTERING THE GEORGIA
6th GRADE CRCT

IN

MATHEMATICS

Developed to the new Georgia Performance Standards!

ERICA DAY

ALAN FUQUA

COLLEEN PINTOZZI

AMERICAN BOOK COMPANY

P. O. BOX 2638

WOODSTOCK, GEORGIA 30188-1383

TOLL FREE 1 (888) 264-5877   PHONE (770) 928-2834

FAX (770) 928-7483

WEB SITE: www.americanbookcompany.com

Acknowledgements

In preparing this book, we would like to acknowledge Mary Stoddard and Eric Field for their contributions in editing and developing graphics for this book. We would also like to thank our many students whose needs and questions inspired us to write this text.

Contents

Acknowledgements ii

Preface viii

Diagnostic Test 1

1 Fractions 14

  1.1   Prime Factorization 14

  1.2   Greatest Common Factor 16

  1.3   Least Common Multiple 17

  1.4   Simplifying Improper Fractions 18

  1.5   Changing Mixed Numbers to Improper Fractions 19

  1.6   Reducing Proper Fractions 20

  1.7   Finding Numerators 21

  1.8   Adding Fractions 22

  1.9   Subtracting Mixed Numbers from Whole Numbers 23

  1.10 Subtracting Mixed Numbers with Borrowing 24

  1.11 Multiplying Fractions 25

  1.12 Dividing Fractions 26

  1.13 Fraction Word Problems 26

  1.14 Comparing the Relative Magnitude of Fractions 27

        Chapter 1 Review 28

2 Decimals 30

  2.1   Adding Decimals 30

  2.2   Subtracting Decimals 31

  2.3   Multiplication of Decimals 32

  2.4   Division of Decimals by Whole Numbers 32

  2.5   Division of Decimals by Decimals 33

  2.6   Changing Fractions to Decimals 34

  2.7   Changing Mixed Numbers to Decimals 35

  2.8   Changing Decimals to Fractions 35

  2.9   Changing Decimals with Whole Numbers to Mixed Numbers 36

  2.10 Decimal Word Problems 36

        Chapter 2 Review 37

3 Percents 38

3.1	Changing Percents to Decimals and Decimals to Percents	38
3.2	Changing Percents to Fractions and Fractions to Percents	39
3.3	Changing Percents to Mixed Numbers and Mixed Numbers to Percents	40
3.4	Comparing the Relative Magnitude of Numbers	41
3.5	Changing to Percent Word Problems	42
3.6	Finding the Percent of the Total	43
3.7	Finding the Percent Increase or Decrease	44
3.8	Tips and Commissions	45
3.9	Finding the Amount of a Discount	46
3.10	Finding the Discounted Sale Price	47
3.11	Sales Tax	48
	Chapter 3 Review	49
4	**Ratios, Proportions, and Scale Drawings**	**50**
4.1	Ratio Problems	50
4.2	Solving Proportions	51
4.3	Ratio and Proportion Word Problems	52
4.4	Proportional Reasoning	53
4.5	Maps and Scale Drawings	54
4.6	Using a Scale On a Blueprint	55
	Chapter 4 Review	56
5	**Patterns and Problem Solving**	**57**
5.1	Number Patterns	57
5.2	Inductive Reasoning and Patterns	58
5.3	Mathematical Reasoning/Logic	62
5.4	Deductive and Inductive Arguments	63
	Chapter 5 Review	65
6	**Solving One-Step Equations**	**66**
6.1	One-Step Algebra Problems with Addition and Subtraction	66
6.2	One-Step Algebra Problems with Multiplication and Division	67
6.3	Multiplying and Dividing with Negative Numbers	68
6.4	Variables with a Coefficient of Negative One	70
	Chapter 6 Review	70
7	**Introduction to Writing and Graphing Equations**	**71**
7.1	Graphing Simple Linear Equations	71

Contents

	7.2	Understanding Slope	73
	7.3	Graphing Linear Data	75
		Chapter 7 Review	77
8	**Data Interpretation**		**78**
	8.1	Tally Charts and Frequency Tables	78
	8.2	Histograms	79
	8.3	Reading Tables	80
	8.4	Bar Graphs	81
	8.5	Line Graphs	82
	8.6	Circle Graphs	83
	8.7	Pictographs	84
	8.8	Graphing Data	86
	8.9	Collecting Data Through Surveys	87
		Chapter 8 Review	87
9	**Probability**		**90**
	9.1	Probability	90
	9.2	More Probability	92
	9.3	Simulations	93
		Chapter 9 Review	94
10	**Measurement**		**96**
	10.1	Using the Ruler	96
	10.2	More Measuring	97
	10.3	Customary Measure	98
	10.4	Approximate English Measure	98
	10.5	The Metric System	99
	10.6	Understanding Meters	99
	10.7	Understanding Liters	99
	10.8	Understanding Grams	99
	10.9	Estimating Metric Measurements	100
	10.10	Converting Units within the Metric System	101
		Chapter 10 Review	102
11	**Plane Geometry**		**103**
	11.1	Perimeter	103
	11.2	Area of Squares and Rectangles	104

11.3 Area of Triangles   105
11.4 Circumference   106
11.5 Area of a Circle   107
11.6 Two-Step Area Problems   107
11.7 Similar and Congruent   110
11.8 Similar Triangles   111
      Chapter 11 Review   113

12 Solid Geometry   **114**
12.1   Understanding Volume   114
12.2   Volume of Rectangular Prisms   115
12.3   Volume of Cubes   116
12.4   Volume of Spheres, Cones, Cylinders, and Pyramids   117
12.5   Two-Step Volume Problems   119
12.6   Geometric Relationships of Solids   120
12.7   Surface Area   122
12.8   Cube   122
12.9   Rectangular Prism   122
12.10 Pyramid   124
12.11 Cylinder   125
12.12 Sphere   126
12.13 Cone   126
12.14 Nets of Solid Objects   127
12.15 Using Nets to Find Surface Area   128
12.16 Solid Geometry Word Problems   130
12.17 Front, Top, Side, and Corner Views of Solid Objects   130
12.18 Compare and Contrast Prisms and Pyramids   134
12.19 Compare and Contrast Cylinders and Cones   135
       Chapter 12 Review   136

13 Symmetry   **138**
13.1 Reflectional Symmetry   138
13.2 Rotational Symmetry   138
13.3 Translational Symmetry   139
13.4 Symmetry Practice   140
      Chapter 13 Review   141

Practice Test 1   **143**

Contents

Practice Test 2   **154**

Index   **163**

Preface

Mastering the Georgia 6th Grade CRCT in Mathematics will help you review and learn important concepts and skills related to middle school mathematics. First, take the Diagnostic Test beginning on page 1 of the book. To help identify which areas are of greater challenge for you, complete the evaluation chart with your instructor in order to help you identify the chapters which require your careful attention. When you have finished your review of all of the material your teacher assigns, take the progress tests to evaluate your understanding of the material presented in this book. **The materials in this book are based on the Georgia Performance Standards including the content descriptions for mathematics, which are published by the Georgia Department of Education. The complete list of standards is located in the Answer Key. Each question in the Diagnostic and Practice Tests is referenced to the standard, as is the beginning of each chapter.**

This book contains several sections. These sections are as follows: 1) A Diagnostic Test; 2) Chapters that teach the concepts and skills for *Mastering the Georgia 6th Grade CRCT in Mathematics*; and 3) Two Practice Tests. Answers to the tests and exercises are in a separate manual.

ABOUT THE AUTHORS

Erica Day is working on a Bachelor of Science Degree in Mathematics at Kennesaw State University, Kennesaw, GA. She is a senior and has been on the Dean's List for her entire undergraduate career. She has also tutored all levels of mathematics, ranging from high school algebra and geometry to university-level statistics and linear algebra. She is currently participating in a mathematics internship for American Book Company, where she does writing and editing.

Alan Fuqua graduated from the Georgia Institute of Technology with a Bachelor of Chemical Engineering degree. He has over fifteen years of industrial experience in the manufacture of inorganic chemicals, including implementing lean manufacturing principles and training employees. He has extensive experience applying statistical models and Six Sigma principles to process improvement and cost savings. He is currently the Mathematics Coordinator for the American Book Company and is continuing his Mathematics education at Kennesaw State University.

Colleen Pintozzi has taught mathematics at the middle school, junior high, senior high, and adult level for 22 years. She hold a B.S. degree from Wright State University in Dayton, Ohio and has done graduate work at Wright State University, Duke University, and the University of North Carolina at Chapel Hill. She is the author of many mathematics books including such best-sellers as *Basics Made Easy: Mathematics Review*, *Passing the New Alabama Graduation Exam in Mathematics*, *Passing the Louisiana LEAP 21 GEE*, *Passing the Indiana ISTEP+ GQE in Mathematics*, *Passing the Minnesota Basic Standards Test in Mathematics,* and *Passing the Nevada High School Proficiency Exam in Mathematics.*

Formula Sheet

Perimeter	Rectangle	$P = 2l + 2w$ or $P = 2(l + w)$
Circumference	Circle	$C = 2\pi r$ or $C = \pi d$
Area	Rectangle	$A = lw$ or $A = bh$
	Triangle	$A = \frac{1}{2}bh$ or $A = \frac{bh}{2}$
	Trapezoid	$A = \frac{1}{2}(b_1 + b_2)h$ or $A = \frac{(b_1 + b_2)h}{2}$
	Circle	$A = \pi r^2$
Surface Area	Cube	$S = 6s^2$
	Cylinder (lateral)	$S = 2\pi rh$
	Cylinder (total)	$S = 2\pi rh + 2\pi r^2$ or $S = 2\pi r(h + r)$
	Cone (lateral)	$S = \pi rl$
	Cone (total)	$S = \pi rl + \pi r^2$ or $S = \pi r(l + r)$
	Sphere	$S = 4\pi r^2$
Volume	Prism or Cylinder	$V = Bh$*
	Pyramid or Cone	$V = \frac{1}{3}Bh$*
	Sphere	$V = \frac{4}{3}\pi r^3$

B represents the area of the Base of a solid figure.

Pi	π	$\pi \approx 3.14$ or $\pi \approx \frac{22}{7}$
Pythagorean Theorem		$a^2 + b^2 = c^2$
Distance Formula		$d = \sqrt{(x_2 - x_1)^2 + (y_2 - y_1)^2}$
Slope of a Line		$m = \frac{y_2 - y_1}{x_2 - x_1}$
Midpoint Formula		$M = (\frac{x_2 + x_1}{2}, \frac{y_2 + y_1}{2})$
Quadratic Formula		$x = \frac{-b \pm \sqrt{b^2 - 4ac}}{2a}$
Slope-Intercept Form of an Equation		$y = mx + b$
Point-Slope Form of an Equation		$y - y_1 = m(x - x_1)$
Standard Form of an Equation		$Ax - By = C$

Diagnostic Test

1. Rami has an aquarium with 3 black goldfish and 4 orange goldfish. He purchased 2 more black goldfish to add to his aquarium. What is the new ratio of black goldfish to total goldfish?

   (A) $\frac{2}{9}$

   (B) $\frac{5}{9}$

   (C) $\frac{4}{5}$

   (D) $\frac{5}{4}$

   M6A1

2. Alice, Barbara and Carol planted a vegetable and flower garden and sold the produce of the garden. They agreed to share the profits as follows: Alice received 30%, Barbara received $\frac{1}{3}$, and Carol received the remainder of the profits. What percent of the profits did Carol receive?

   (A) $33\frac{1}{3}\%$
   (B) $36\frac{2}{3}\%$
   (C) 40%
   (D) $69\frac{2}{3}\%$

   M6N1g

3. The number of various kinds of animals treated by three veterinarians is shown in the table below.

	Dr. Ortez	Dr. Smith	Dr. Wood
Dogs	13	7	12
Cats	5	7	9
Horses	3	1	0
Other	4	2	2

   Of all the animals treated, the percent that is not cats or dogs is most nearly

   (A) 6%
   (B) 12%
   (C) 18%
   (D) 20%

   M6N1g

4. Monique is examining a scale model of an ancient building as a part of her research for a history project. If the scale is $\frac{1}{4}$ inch to one foot and the model is 24 inches long, what is the actual length of the building?

   (A) 6 feet
   (B) 20 feet
   (C) 28 feet
   (D) 96 feet

   M6G1e

5. $\frac{5}{8}$ written as a percent is

   (A) 0.58%
   (B) 0.625%
   (C) 6.25%
   (D) 62.5%

   M6N1f

6. Last year, there were 96 students in the marching band. This year, the band's size has increased by 25%. How many students are in the marching band this year?

   (A) 121
   (B) 120
   (C) 24
   (D) 125

   M6N1g

7. What is the value of $\frac{5}{7} + \frac{1}{3}$

   (A) $\frac{6}{10}$

   (B) $\frac{6}{7}$

   (C) $\frac{22}{7}$

   (D) $\frac{22}{21}$

   M6N1d

1

8. Below is a drawing of a farm plot:

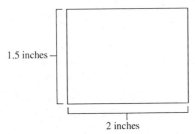

1.5 inches

2 inches

Scale: 1 inch = 0.75 miles

What is the perimeter of this farm plot?

(A) $2\frac{1}{4}$ miles
(B) 3 miles
(C) $5\frac{1}{4}$ miles
(D) 7 miles

M6C1d

9. What is the area of the shaded region below?

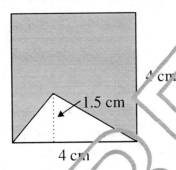

4 cm
1.5 cm
4 cm

(A) 16 cm²
(B) 13 cm²
(C) 10 cm²
(D) 6 cm²

M6M3b

10. Find the volume of the pyramid.

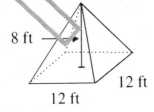

8 ft
12 ft
12 ft

(A) 384 ft³
(B) 36 ft³
(C) 32 ft³
(D) 1 ft³

M6M3b

11. Sarah has $11.27, and buys a movie for $5.50. How much money does she have left?

(A) $16.77
(B) $5.50
(C) $5.00
(D) $2.05

M6N1g

12. What is the value of $\frac{4}{3} - \frac{2}{5}$?

(A) -1
(B) $\frac{14}{15}$
(C) $\frac{2}{5}$
(D) $\frac{8}{15}$

M6N1d

13. Mrs. Campbell's 6th grade class is going on a field trip. There are 29 children in the class. Parents are driving, and there will be 4 students per car. What is the smallest number of cars they will need for the children?

(A) 6
(B) 7
(C) 8
(D) 9

M6A2g

14. Melanie put her car in cruise control and noted her progress every 15 minutes.

Time	Distance Traveled
15 min	18.5 miles
30 min	37 miles
45 min	55.5 miles
60 min	74 miles
75 min	92.5 miles

Assuming the pattern continues, how far will she travel in 2 hours 45 minutes?

(A) 129.5 miles
(B) 193 miles
(C) 203.5 miles
(D) 283.5 miles

M6A2a

2

15. Janice bought 10 yards of fabric to recover 6 dining room chairs. Each chair took $1\frac{1}{4}$ yards. How much fabric did she have left?

(A) $2\frac{1}{2}$

(B) 4

(C) $4\frac{4}{5}$

(D) $7\frac{1}{2}$

M6N1e

16. $\dfrac{3}{8}$ written as a decimal is

(A) 0.375
(B) 0.38
(C) 0.0375
(D) 0.3

M6N1f

17. What is the volume of a cube that is 7 inches on each edge?

(A) 42 in^3
(B) 49 in^3
(C) 84 in^3
(D) 343 in^3

M6M3b

18. In the figures below, an edge of the larger cube is twice as big as an edge of the smaller cube. What is the ratio of the volume of the smaller cube to that of the larger cube?

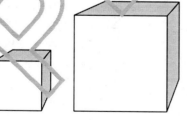

(A) 1:2
(B) 1:4
(C) 1:8
(D) 1:16

M6M3d

19. Consider the solid shown below. Which of the following diagrams represents the side view of this object?

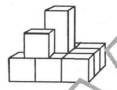

(A)

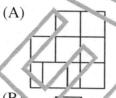

(B)

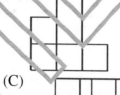

(C)

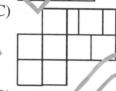

(D)

M6G2c

20.

The capital letter A shown above has what kind of symmetry?

(A) line symmetry
(B) point symmetry
(C) rotational symmetry
(D) none

M6G1a

3

21. Which of the following Cartesian planes is an accurate graph of the point values below?

Cups	Ounces
1	8
2	16
3	24

(A)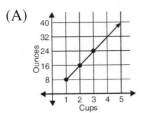

(B)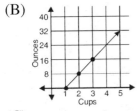

(C)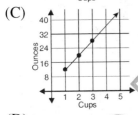

(D)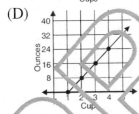

M6A2e

22. Justin sells hats at the July 4th parade. He buys the hats wholesale for $3 each, marks up 50%, and then adds 10% to cover tax and license fees. What price does he charge for each hat?

(A) $3.60
(B) $4.50
(C) $4.65
(D) $4.95

M6N1g

23. Central High School had a dance last Saturday night. The graph below shows total number of students in each grade and the number of students in each grade that attended the dance.

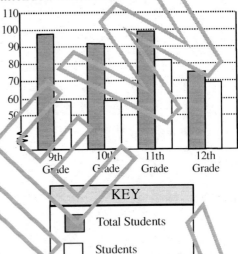

Which grade had the highest proportion of students in that grade attending the dance?

(A) 9th grade
(B) 10th grade
(C) 11th grade
(D) 12th grade

M6A2c

24. Jamie deposits $0.50 into Miss Clucky, a machine that makes chicken squawks, and it gives Jamie one plastic egg with a toy surprise. In the machine, 30 eggs contain a rubber frog, 43 eggs contain a plastic ring, 23 eggs contain a necklace, and 18 eggs contain a plastic car. What is the probability that Miss Clucky will give Jamie a necklace in her egg?

(A) $\frac{1}{114}$

(B) $\frac{23}{114}$

(C) $\frac{23}{91}$

(D) $\frac{1}{23}$

M6D2b

4

25. What is the length of the line segment $\overline{WY}$?

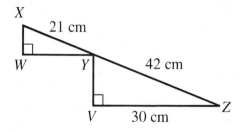

(A) 15 cm
(B) 16 cm
(C) 18 cm
(D) 30 cm

M6G1c

26. Amin has a part-time job earning $10.00 per hour. He made a chart of his hours, earnings, and federal taxes taken out of his paycheck.

Hours	Earned	Taxes
25	$250	$23
26	$260	$25
27	$270	$26
28	$280	$28
29	$290	$29

If the pattern continues, how much will be taken from his check for federal taxes if he works 32 hours?

(A) $32
(B) $33
(C) $34
(D) $35

M6A2a

27. If you want to show the continuous growth of a population over time, which type of graph would you use?

(A) line graph
(B) circle graph
(C) histogram
(D) bar graph

M6D1c

28. Find c: $\dfrac{c}{-2} = -6$

(A) $c = -12$
(B) $c = 12$
(C) $c = -3$
(D) $c = 3$

M6A3

29. Tom's school is considering making uniforms mandatory starting with the next school year. Tom hates the idea and wants to do his own survey to see if parents are really in favor of it. He considers 4 places to conduct his survey. Which would give the most valid results?

(A) He would stop people at random walking through the mall.
(B) He would survey parents in the car pool lanes picking up students after school.
(C) He would survey the teachers after school.
(D) He would survey the students in his biology class to ask what their parents thought.

M6D1a

30. Which of the following is the prime factorization of 40?

(A) $2^3 \times 5$
(B) $3^2 \times 5$
(C) 4×10
(D) $2 \times 5 \times 4$

M6N1b

31. The area of a picture frame is 2 square feet. How many square inches is that?

(A) 288 square inches
(B) 24 square inches
(C) 48 square inches
(D) 96 square inches

M6M1

32. The scale drawing of an advertising sign is drawn with a scale of 1 inch = 4 feet.

$5\frac{7}{16}$"

$3\frac{5}{8}$"

Now Open For Business

What is the actual width of the advertising sign?

(A) $5\frac{3}{8}$ feet
(B) $13\frac{1}{2}$ feet
(C) $14\frac{1}{2}$ feet
(D) 58 feet   M6G1e

33. Emily needs to make a glass case with the following measurements:

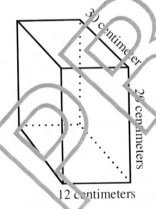

30 centimeter
20 centimeters
12 centimeters

How many square centimeters of glass would it take to construct the case enclosed on all sides?

(A) 60 square centimeters
(B) 612 square centimeters
(C) 2,400 square centimeters
(D) 6,200 square centimeters   M6M4b

34. Cynda wants to buy a day-care center with the measurements below. How many square feet are in the building?

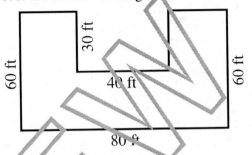

60 ft
30 ft
40 ft
60 ft
80 ft

(A) 540 square feet
(B) 420 square feet
(C) 3,600 square feet
(D) 4,800 square feet   M6M2b

35. Find the volume of the figure below.

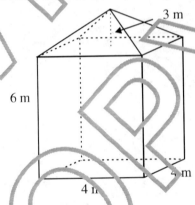

3 m
6 m
3 m
4 m

(A) 96 m³
(B) 99 m³
(C) 112 m³
(D) 288 m³   M6M3b

36. In Betty's class there are 16 girls and 14 boys. Which of these is the correct ratio of girls to the total number of students in the class?

(A) 14 to 16
(B) 16 to 14
(C) 14 to 30
(D) 16 to 30   M6A1

6   Copyright ©American Book Company

37. Which of the following nets represents a cube?

(A)

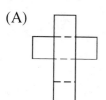

(B)

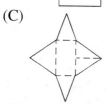

(C)

(D)

M6G2d

38. A recipe for 32 ounces of lemonade calls for 4 ounces of lemon juice. Allie wants to make 120 ounces of lemonade. Which proportion below should she use to find the amount of lemon juice needed?

(A) $\frac{32}{120} = \frac{x}{4}$

(B) $\frac{x}{32} = \frac{4}{120}$

(C) $\frac{32}{4} = \frac{x}{120}$

(D) $\frac{4}{32} = \frac{x}{120}$

M6A2b

39. Kevin counted the number of computers in 53 classrooms. The bar graph shows the results of his count.

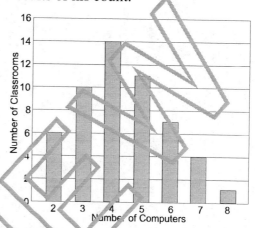

How many classrooms had less than 5 computers?

(A) 14
(B) 15
(C) 20
(D) 30

M6D1e

40. The school counselor asked 60 students, "What is your favorite class?" The results are shown in the circle graph (pie chart). How many students said "Lunch" was their favorite subject?

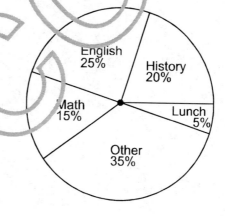

(A) 3
(B) 5
(C) 8
(D) 18

M6D1e

7

41. Elisha set out walking to the bus stop. Suddenly, she realized she had forgotten her lunch box. She ran back home, found her lunch box, and ran to the bus stop so as not to miss her bus. Which of the following graphs best models this situation?

(A)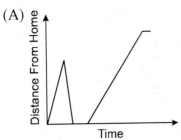

(B)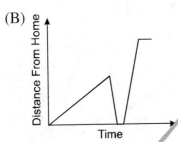

(C)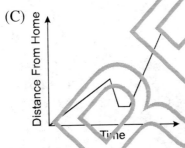

(D)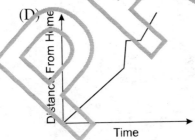

M6D1c

42. 2 kiloliters is equal to:

(A) 2000 liters
(B) 200 liters
(C) 20 liters
(D) .002 liters

M6M1

43. $\triangle ABC$ is similar to $\triangle DEF$. What is the value of x?

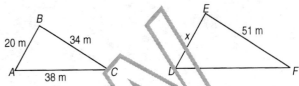

Note: The figures are not drawn to scale.

(A) 17 m
(B) 22 m
(C) 30 m
(D) 37 m

M6G1c

44. On a blueprint of a house, the scale is 0.25 inches equals 2 feet. How wide is the kitchen if it measures 1.5 inches on the blueprint?

(A) $0.\overline{33}$ feet
(B) 3 feet
(C) 12 feet
(D) 15 feet

M6G1e

45. What would be the best measure to use to find the area of a school lunchroom?

(A) square kilometers
(B) square meters
(C) square centimeters
(D) square millimeters

M6M2c

46. Which of the following lists all the factors of the number 12?

(A) 1, 12
(B) 2, 3, 4, 6
(C) 5, 11
(D) none of the above

M6N1a

47. 122% written as a fraction is

(A) $1\frac{11}{50}$

(B) $2\frac{3}{25}$

(C) $12\frac{1}{5}$

(D) 122   M6N1f

48. Which of the following is equal to 30 millimeters?

(A) 0.03 meters
(B) 0.003 kilometers
(C) 300 centimeters
(D) 30,000 meters   M6M1

49. Which of the following is equal to 5 L?

(A) 0.005 mL
(B) 0.5 kL
(C) 5,000 mL
(D) 5,000 kL   M6M1

50. The Thompson family wants to make a graph to show the percent of money spent on vacation in the following categories:

entertainment   5%
hotel   55%
souvenirs   8%
food   20%
gasoline   12%

What is the best kind of graph to use to display their results?

(A) bar graph
(B) single line graph
(C) circle graph
(D) multiple line graph   M6D1c

51. If $\frac{n}{6} = -8$, what is n?

(A) -48
(B) 14
(C) 8
(D) -2 M6A3

52. What is the surface area of a cylinder with an 8 inch diameter and a height of 3 inches? Use the formula $SA = 2\pi r^2 + 2\pi rh$ Use $\pi = 3.14$

(A) 100.48 in^2
(B) 50.24 in^2
(C) 75.36 in^2
(D) 175.84 in^2 M6M4b

53. A rectangular box and a rectangular pyramid have the same dimensions for their bases and heights. How does the volume of the box compare to the volume of the pyramid?

(A) The volumes are the same.
(B) The volume of the box is twice the volume of the pyramid.
(C) The volume of the box is three times as large as the pyramid.
(D) The volume of the box is four times as large as the pyramid.   M6G2a

54. Jack is going to paint the ceiling and four walls of a room that is 10 feet wide, 12 feet long, and 10 feet from floor to ceiling. How many square feet will he paint?

(A) 120 square feet
(B) 560 square feet
(C) 680 square feet
(D) 1,200 square feet   M6M4d

55. What kind of symmetry does the following figure have? Choose the best answer.

I. reflectional symmetry
II. 90° rotational symmetry
III translational symmetry

(A) I
(B) II
(C) I and II
(D) III

M6G1b

56.

The circle graph most accurately represents which of the situations below?

(A) John's after-school activities between 4:00 and 6:00 p.m. are divided between watching TV, 30%; playing video games, 20%; playing sports, 30%; and surfing the internet, 20%.
(B) The Johnson household family budget is divided into the following categories: housing, 40%; food, 25%; clothing, 20%; charities, 15%.
(C) The basketball team spends their budget in the following categories: uniforms, 50%; equipment, 10%; travel, 30%; and snacks, 10%.
(D) Tina spends her $80.00 monthly allowance in the following categories: make-up, 25%; clothes, 50%; snacks, 5%; and music CDs, 20%.

M6D1c

57. How many of the smaller cubes will fit inside the larger box? (Figures are not drawn to scale.)

 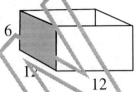

(A) 6
(B) 16
(C) 20
(D) 32

M6M3d

58. Casey, our pet iguana, whacked his tail into a pile of marbles. One of the marbles went under the couch, never to be seen again. There were 6 red marbles, 11 orange marbles, 4 blue marbles, and 7 multicolored marbles. What is the probability the one missing is not orange?

(A) 11 out of 28
(B) 1 out of 28
(C) 17 out of 28
(D) 27 out of 28

M6D2b

59. A net for a rectangular prism is shown below. What is the surface area of the prism?

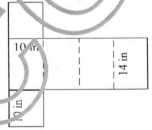

(A) 680 in²
(B) 760 in²
(C) 800 in²
(D) 840 in²

M6M4a

60. Mrs. Jackson has 4 dry-erase markers; they are blue, red, green and orange. Each time she uses the dry-erase board Mrs. Jackson randomly chooses one of the four dry-erase markers. Last week, of the 16 times she used the dry-erase board, she chose a green marker 3 times. How does the theoretical probability of choosing a green marker compare to Mrs. Jackson's experimental probability?

(A) The theoretical probability is 6.25% lower than the experimental probability.
(B) The theoretical probability is 6.25% higher than the experimental probability.
(C) The theoretical and experimental probabilities are equal.
(D) Not enough information was given in the problem to calculate the theoretical and experimental probabilities. M6D2a

61. Lorie uses 510 cm³ of sand to fill a cube. How many cubic centimeters of sand would be used to fill a pyramid of the same height and base as the cube?

(A) 170 cm³
(B) 255 cm³
(C) $127\frac{1}{2}$ cm³
(D) 510 cm³ M6G2a

62. Which of the following sets contains equivalent numbers?

(A) $\frac{9}{25}$ 0.35 35%

(B) $\frac{5}{16}$ 0.315 $31\frac{1}{2}\%$

(C) $\frac{3}{8}$ 0.375 $37\frac{1}{2}\%$

(D) $\frac{4}{5}$ 0.08 80% M6N1f

63. $2\frac{1}{2} \times 3\frac{5}{9} \times \frac{3}{5}$

(A) $3\frac{5}{6}$

(B) $5\frac{1}{3}$

(C) $5\frac{2}{3}$

(D) $6\frac{1}{6}$ M6N1e

64. $1\frac{3}{4} \div 2\frac{5}{8}$

(A) $\frac{1}{2}$

(B) $\frac{2}{3}$

(C) $\frac{7}{8}$

(D) $1\frac{1}{2}$ M6N1e

65. Find the multiple of 3, 11 and 22 that is less than 100.

(A) 25
(B) 36
(C) 66
(D) 99 M6N1a

66. What is the prime factorization of 72?

(A) 8×9
(B) 1×72
(C) 2×6^2
(D) $2^3 \times 3^2$ M6N1b

67. For every 4 fish that Alice has in her pond, she must have five plants for them. If she only has 75 plants, what is the total number of fish she can have?

(A) 60
(B) 94
(C) 135
(D) 169

M6A2g

68. Which of the following sets of numbers include a number which is not a prime number?

(A) 23   29   57
(B) 2   53   59
(C) 31   47   61
(D) 5   13   73

M6N1b

69. What is the GCF of 14 and 6?

(A) 1
(B) 2
(C) 28
(D) 42

M6N1c

70. What is the LCM of 12 and 8?

(A) 2
(B) 4
(C) 8
(D) 24

M6N1c

Evaluation Chart for the Diagnostic Mathematics Test

Directions: On the following chart, circle the question numbers that you answered incorrectly. Then turn to the appropriate topics, read the explanations, and complete the exercises. Review the other chapters as needed. Finally, complete the *Mastering the Georgia 6th Grade CRCT in Mathematics* Practice Tests to further review.

		Questions	Pages
Chapter 1:	Fractions	7, 12, 15, 30, 46, 63, 64, 65, 66, 68, 69, 70	14–29
Chapter 2:	Decimals	11, 16	30–37
Chapter 3:	Percents	2, 5, 6, 22, 47, 62	38–49
Chapter 4:	Ratios, Proportions, and Scale Drawings	1, 4, 13, 32, 36, 38, 44, 67	50–56
Chapter 5:	Patterns and Problem Solving	14, 26	57–65
Chapter 6:	One-Step Equations	28, 51	66–70
Chapter 7:	Introduction to Writing and Graphing Equations	21	71–77
Chapter 8:	Data Interpretation	3, 23, 27, 29, 39, 40, 41, 50, 56	78–89
Chapter 9:	Probability	24, 58, 60	90–95
Chapter 10:	Measurement	31, 42, 45, 48, 49	96–102
Chapter 11:	Plane Geometry	8, 9, 25, 34, 43	103–113
Chapter 12:	Solid Geometry	10, 17, 18, 19, 33, 35, 37, 52, 53, 54, 57, 59, 61	114–137
Chapter 13:	Symmetry	20, 55	138–141

Chapter 1
Fractions

This chapter covers the following Georgia Performance Standards:

M6N	Number and Operations	M6N1.a, b, c, d, e, f, g
M6P	Process Skills	M6P1.a, b, c, d
		M6P3.a, c, d
		M6P4.e
		M6P5.a, b, c

1.1 Prime Factorization

Prime factorization is the process of factoring a number into prime numbers. A prime number, also called a prime, is a number that can only be divided by itself and 1. There are two main ways of finding the primes of a number: dividing and splitting.

Example 1: Find the primes of 66 by division.

   Step 1: To find the primes by division, you must only divide 66 by prime numbers until you can only divide by one.
$66 \div 2 = 33, 33 \div 3 = 11, 11 \div 11 = 1, 11 \div 1 = 11$ (1 is not prime)

   Step 2: All the prime numbers used as divisors make up the prime factorization of 6.
$66 = 2 \times 3 \times 11$

   Check: To check, multiply the prime numbers together, and you should get the original value.

Example 2: Find the primes of 66 and 120 using the splitting method.

   Step 1: In this method, the number must be split by any two factors until all of the factors are prime.

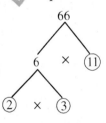

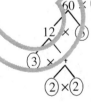

$66 = 2 \times 3 \times 11$ $\qquad$ $120 = 2 \times 2 \times 2 \times 5 \times 3 = 2^3 \times 3 \times 5$

   Step 2: All the prime numbers used in splitting 66 make up the prime factorization of 66. All the prime numbers used in splitting 120 make up the prime factorization of 120.

   Hint: The factors found during prime factorization should always be prime, should always multiply together to get the correct answer, and should always be listed from least to greatest.

Find the prime factorization of each number using the division method. The first one has been done for you.

1. $10 \div 2 = 5 \div 5 = 1$
   $10 = 2 \times 5$

2. 14

3. 55

4. 110

5. 126

6. 142

7. 8

8. 21

9. 32

10. 36

11. 51

12. 84

13. 125

14. 48

15. 77

16. 65

17. 200

18. 413

Find the prime factorization of each number using the splitting method. The first one has been done for you.

19.

$12 = 2 \times 2 \times 3 = 2^2 \times 3$

20. 24

21. 45

22. 120

23. 52

24. 91

25. 18

26. 67

27. 20

28. 15

29. 35

30. 122

1.2   Greatest Common Factor

To reduce to their simplest form, you must be able to find the greatest common factor.

Example 3:   Find the greatest common factor (GCF) of 16 and 24.

To find the **greatest common factor (GCF)** of two numbers, first list the factors of each number. The factors are all the numbers that will divide evenly into the numbers that you want to find the factors for.

The factors of 16 are: 1, 2, 4, 8, and 16

The factors of 24 are: 1, 2, 3, 4, 5, 8, 12, and 24

What is the **largest** number they both have in common? **8**

8 is the **greatest** (largest number) **common factor**.

Find all the factors and the greatest common factor (GCF) of each pair of numbers below.

	Pairs	Factors	GCF		Pairs	Factors	GCF
1.	10			10.	6		
	15				42		
2.	12			11.	14		
	16				63		
3.	18			12.	9		
	36				51		
4.	27			13.	18		
	45				45		
5.	32			14.	12		
	40				20		
6.	16			15.	16		
	48				40		
7.	14			16.	10		
	42				45		
8.	4			17.	18		
	26				30		
9.	8			18.	15		
	28				25		

1.3 Least Common Multiple

To find the **least common multiple (LCM)**, of two numbers, first list the multiples of each number. The multiples of a number are 1 times the number, 2 times the number, 3 times the number, and so on.

The multiples of 6 are: 6, 12, 18, 24, 30...

The multiples of 10 are: 10, 20, 30, 40, 50...

What is the smallest multiple they both have in common?   30

30 is the **least** (smallest number) **common multiple** of 6 and 10.

Find the least common multiple (LCM) of each pair of numbers below.

	Pairs	Multiples	LCM		Pairs	Multiples	LCM
1.	6	6, 12, 18, 24, 30	30	10.	6		
	15	15, 30			7		
2.	12			11.	4		
	16				18		
3.	18			12.	7		
	36				5		
4.	7			13.	30		
	3				45		
5.	12			14.	3		
	8				8		
6.	6			15.	12		
	8				9		
7.	4			16.	5		
	14				45		
8.	9			17.	3		
	6				5		
9.	2			18.	4		
	15				22		

1.4   Simplifying Improper Fractions

Example 4:   Simplify $\dfrac{21}{4} = 21 \div 4 = 5$ remainder 1

The quotient, 5, becomes the whole number portion of the mixed number.

$$\dfrac{21}{4} = 5\dfrac{1}{4}$$ ⇂ The remainder, 1, becomes the top number of the fraction.

The bottom number of the fraction always remains the same.

Example 5:   Simplify $\dfrac{11}{6}$.

Step 1: $\dfrac{11}{6}$ is the same as $11 \div 6$. $11 \div 6 = 1$ with a remainder of 5.

Step 2:   Rewrite as a whole number with a fraction. $1\dfrac{5}{6}$

Simplify the following improper fractions.

1. $\dfrac{13}{5} =$ _____   5. $\dfrac{19}{6} =$ _____   9. $\dfrac{22}{3} =$ _____   13. $\dfrac{17}{9} =$ _____   17. $\dfrac{7}{4} =$ _____

2. $\dfrac{11}{3} =$ _____   6. $\dfrac{16}{7} =$ _____   10. $\dfrac{13}{4} =$ _____   14. $\dfrac{27}{8} =$ _____   18. $\dfrac{21}{10} =$ _____

3. $\dfrac{24}{6} =$ _____   7. $\dfrac{13}{8} =$ _____   11. $\dfrac{15}{2} =$ _____   15. $\dfrac{32}{7} =$ _____

4. $\dfrac{7}{6} =$ _____   8. $\dfrac{9}{5} =$ _____   12. $\dfrac{22}{9} =$ _____   16. $\dfrac{3}{2} =$ _____

Fractions that have the same denominator (bottom number) can be added quickly. Add the numerators (top numbers) and keep the bottom number the same. Simplify the answer. The first one is done for you.

19.
$$\begin{array}{r} \dfrac{2}{9} \\ \dfrac{7}{9} \\ +\dfrac{4}{9} \\ \hline \dfrac{13}{9} = 1\dfrac{4}{9} \end{array}$$

20.
$$\begin{array}{r} \dfrac{2}{6} \\ \dfrac{4}{6} \\ +\dfrac{5}{6} \\ \hline \end{array}$$

21.
$$\begin{array}{r} \dfrac{3}{8} \\ \dfrac{5}{8} \\ +\dfrac{7}{8} \\ \hline \end{array}$$

22.
$$\begin{array}{r} \dfrac{9}{10} \\ \dfrac{1}{10} \\ +\dfrac{3}{10} \\ \hline \end{array}$$

23.
$$\begin{array}{r} \dfrac{8}{13} \\ \dfrac{6}{13} \\ +\dfrac{2}{13} \\ \hline \end{array}$$

24.
$$\begin{array}{r} \dfrac{6}{7} \\ \dfrac{4}{7} \\ +\dfrac{5}{7} \\ \hline \end{array}$$

25.
$$\begin{array}{r} \dfrac{3}{5} \\ \dfrac{4}{5} \\ +\dfrac{5}{7} \\ \hline \end{array}$$

1.5   Changing Mixed Numbers to Improper Fractions

Example 6:   Change $4\frac{3}{5}$ to an improper fraction.

Step 1:   Multiply the whole number (4) by the bottom number of the fraction (5).
$4 \times 5 = 20$

Step 2:   Add the top number to the product from Step 1. $20 + 3 = 23$

Step 3:   Put the answer over the bottom number (5).

2. Add this number.   3. Put the answer here.

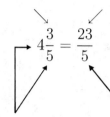

$$4\frac{3}{5} = \frac{23}{5}$$

4. This number stays the same.

1. Multiply these two numbers.

Change the following mixed numbers to improper fractions.

1. $3\frac{1}{2} =$ _____

2. $2\frac{7}{8} =$ _____

3. $9\frac{2}{3} =$ _____

4. $4\frac{3}{5} =$ _____

5. $7\frac{1}{4} =$ _____

6. $8\frac{5}{8} =$ _____

7. $1\frac{2}{7} =$ _____

8. $2\frac{4}{9} =$ _____

9. $6\frac{1}{5} =$ _____

10. $5\frac{2}{7} =$ _____

11. $3\frac{3}{5} =$ _____

12. $9\frac{3}{8} =$ _____

13. $10\frac{4}{5} =$ _____

14. $3\frac{3}{10} =$ _____

15. $4\frac{1}{7} =$ _____

16. $2\frac{5}{6} =$ _____

17. $7\frac{3}{7} =$ _____

18. $6\frac{7}{9} =$ _____

19. $7\frac{2}{5} =$ _____

20. $1\frac{6}{7} =$ _____

Whole numbers become improper fractions when you put them over 1. The first one is done for you.

21. $4 = \frac{4}{1}$

22. $10 =$ _____

23. $3 =$ _____

24. $2 =$ _____

25. $15 =$ _____

26. $5 =$ _____

27. $6 =$ _____

28. $11 =$ _____

29. $8 =$ _____

30. $16 =$ _____

1.6   Reducing Proper Fractions

Example 7:   Reduce $\frac{4}{8}$ to lowest terms.

Step 1:   First you need to find the greatest common factor of 4 and 8. Think: What is the largest number that can be divided into 4 and 8 without a remainder?

These must be the same number. $\searrow$ $?\overline{)4}$ 4 and 8 can both be divided by 4.
$\nearrow$ $?\overline{)8}$

Step 2:   Divide the top and bottom of the fraction by the same number.
$\frac{4 \div 4}{8 \div 4} = \frac{1}{2}$ Therefore, $\frac{4}{8} = \frac{1}{2}$.

Reduce the following fraction to lowest terms.

1. $\frac{2}{8}$

2. $\frac{12}{15}$

3. $\frac{9}{27}$

4. $\frac{12}{42}$

5. $\frac{3}{21}$

6. $\frac{27}{54}$

7. $\frac{14}{22}$

8. $\frac{9}{21}$

9. $\frac{4}{14}$

10. $\frac{6}{26}$

11. $\frac{30}{45}$

12. $\frac{16}{64}$

13. $\frac{10}{25}$

14. $\frac{3}{12}$

15. $\frac{15}{30}$

16. $\frac{12}{36}$

17. $\frac{13}{39}$

18. $\frac{28}{49}$

19. $\frac{8}{18}$

20. $\frac{14}{21}$

21. $\frac{2}{12}$

22. $\frac{5}{15}$

23. $\frac{9}{15}$

24. $\frac{24}{48}$

25. $\frac{3}{18}$

26. $\frac{6}{27}$

27. $\frac{4}{18}$

28. $\frac{8}{28}$

29. $\frac{14}{42}$

30. $\frac{18}{36}$

1.7   Finding Numerators

Remember, any fraction that has the same non-zero numerator (top numbers) and denominator (bottom number) equals 1.

Example 8: $\dfrac{5}{5} = 1$ $\dfrac{8}{8} = 1$ $\dfrac{12}{12} = 1$ $\dfrac{15}{15} = 1$ $\dfrac{25}{25} = 1$

Any fraction multiplied by 1 in any form remains equal to itself.

Example 9: $\dfrac{3}{7} \times \dfrac{4}{4} = \dfrac{12}{28}$ so $\dfrac{3}{7} = \dfrac{12}{28}$

Find the missing numerator (top number) $\dfrac{5}{8} = \dfrac{}{24}$

Step 1:   Ask yourself, "What was 8 multiplied by to get 24?"   3 is the answer.

Step 2:   The only way to keep the fraction equal is to multiply the top and bottom number by the same number. The bottom number was multiplied by 3. so multiply the top number by 3 as shown below.

$$\dfrac{5}{8} \times \dfrac{3}{3} = \dfrac{15}{24}$$

Find the missing numerators from the following equivalent fractions.

1. $\dfrac{2}{6} = \dfrac{}{18}$

2. $\dfrac{2}{3} = \dfrac{}{27}$

3. $\dfrac{4}{9} = \dfrac{}{18}$

4. $\dfrac{7}{15} = \dfrac{}{45}$

5. $\dfrac{9}{10} = \dfrac{}{50}$

6. $\dfrac{5}{6} = \dfrac{}{36}$

7. $\dfrac{1}{4} = \dfrac{}{36}$

8. $\dfrac{3}{14} = \dfrac{}{28}$

9. $\dfrac{2}{5} = \dfrac{}{25}$

10. $\dfrac{4}{11} = \dfrac{}{33}$

11. $\dfrac{5}{6} = \dfrac{}{18}$

12. $\dfrac{6}{11} = \dfrac{}{22}$

13. $\dfrac{8}{15} = \dfrac{}{45}$

14. $\dfrac{1}{9} = \dfrac{}{18}$

15. $\dfrac{7}{8} = \dfrac{}{40}$

16. $\dfrac{1}{12} = \dfrac{}{48}$

17. $\dfrac{3}{8} = \dfrac{}{24}$

18. $\dfrac{3}{4} = \dfrac{}{16}$

19. $\dfrac{2}{7} = \dfrac{}{49}$

20. $\dfrac{11}{12} = \dfrac{}{24}$

21. $\dfrac{2}{5} = \dfrac{}{45}$

22. $\dfrac{4}{5} = \dfrac{}{15}$

23. $\dfrac{1}{9} = \dfrac{}{27}$

24. $\dfrac{3}{8} = \dfrac{}{56}$

25. $\dfrac{3}{13} = \dfrac{}{26}$

26. $\dfrac{1}{7} = \dfrac{}{35}$

27. $\dfrac{4}{5} = \dfrac{}{10}$

28. $\dfrac{3}{10} = \dfrac{}{40}$

29. $\dfrac{7}{8} = \dfrac{}{48}$

30. $\dfrac{6}{7} = \dfrac{}{14}$

1.8   Adding Fractions

Example 10:   Add $3\frac{1}{2} + 2\frac{2}{3}$

Step 1:   Rewrite the problem vertically, and find a common denominator. Think: What is the smallest number I can divide 2 and 3 into without a remainder? 6, of course.

$$3\frac{1}{2} = \frac{}{6}$$
$$+2\frac{2}{3} = \frac{}{6}$$

Step 2:   To find the numerator for the top fraction, think. What do I multiply 2 by to get 6? You must multiply the top and bottom numbers of the fraction by 3 to keep the fraction equal. For the bottom fraction, multiply the top and bottom number by 2.

Step 3:   Add whole numbers and fractions, and simplify.

$$3\frac{1}{2} = 3\frac{3}{6}$$
$$+2\frac{2}{3} = 2\frac{4}{6}$$
$$= 5\frac{7}{6} = 6\frac{1}{6}$$

Add and simplify the answers.

1. $3\frac{5}{9}$
   $+5\frac{2}{3}$

2. $1\frac{1}{4}$
   $+4\frac{2}{5}$

3. $3\frac{3}{4}$
   $+2\frac{3}{5}$

4. $2\frac{1}{4}$
   $+1\frac{7}{8}$

5. $6\frac{5}{6}$
   $+4\frac{1}{3}$

6. $9\frac{1}{5}$
   $+5\frac{5}{6}$

7. $1\frac{1}{3}$
   $+7\frac{3}{4}$

8. $9\frac{4}{9}$
   $+3\frac{2}{3}$

9. $1\frac{7}{10}$
   $+8\frac{2}{3}$

10. $5\frac{2}{7}$
    $+\frac{1}{2}$

11. $3\frac{3}{11}$
    $+2\frac{3}{4}$

12. $\frac{3}{5}$
    $+\frac{4}{9}$

1.9   Subtracting Mixed Numbers from Whole Numbers

Example 11:   Subtract $15 - 3\frac{3}{4}$

Step 1:   Rewrite the problem vertically.

$$\begin{array}{r} 15 \\ - \quad 3\frac{3}{4} \\ \hline \end{array}$$

Step 2:   You cannot subtract three-fourths from nothing. You must borrow 1 from 15. You will need to put the 1 in the fraction form. If you use $\frac{4}{4}$ $\left(\frac{4}{4} = 1\right)$, you will be ready to subtract.

$$\begin{array}{r} 4 \\ 1\overset{\;}{5}\frac{4}{4} \\ - \quad 3\frac{3}{4} \\ \hline 11\frac{1}{4} \end{array}$$

Subtract.

1. $\begin{array}{r} 12 \\ - \quad 3\frac{2}{9} \\ \hline \end{array}$

2. $\begin{array}{r} 3 \\ - \quad 1\frac{4}{7} \\ \hline \end{array}$

3. $\begin{array}{r} 24 \\ - \quad 11\frac{4}{5} \\ \hline \end{array}$

4. $\begin{array}{r} 2 \\ - \quad 1\frac{2}{5} \\ \hline \end{array}$

5. $\begin{array}{r} 4 \\ - \quad 1\frac{5}{8} \\ \hline \end{array}$

6. $\begin{array}{r} 11 \\ - \quad 9\frac{7}{8} \\ \hline \end{array}$

7. $\begin{array}{r} 14 \\ - \quad 9\frac{7}{12} \\ \hline \end{array}$

8. $\begin{array}{r} 8 \\ - \quad 3\frac{1}{3} \\ \hline \end{array}$

9. $\begin{array}{r} 5 \\ - \quad 3\frac{1}{2} \\ \hline \end{array}$

10. $\begin{array}{r} 17 \\ - \quad 13\frac{1}{5} \\ \hline \end{array}$

11. $\begin{array}{r} 3 \\ - \quad 1\frac{5}{11} \\ \hline \end{array}$

12. $\begin{array}{r} 13 \\ - \quad 8\frac{9}{10} \\ \hline \end{array}$

13. $\begin{array}{r} 15 \\ - \quad 6\frac{3}{4} \\ \hline \end{array}$

14. $\begin{array}{r} 6 \\ - \quad 4\frac{8}{9} \\ \hline \end{array}$

15. $\begin{array}{r} 20 \\ - \quad 12\frac{6}{7} \\ \hline \end{array}$

16. $\begin{array}{r} 21 \\ - \quad 1\frac{3}{20} \\ \hline \end{array}$

17. $\begin{array}{r} 9 \\ - \quad 5\frac{2}{3} \\ \hline \end{array}$

18. $\begin{array}{r} 8 \\ - \quad 7\frac{3}{5} \\ \hline \end{array}$

19. $\begin{array}{r} 5 \\ - \quad 4\frac{5}{8} \\ \hline \end{array}$

20. $\begin{array}{r} 14 \\ - \quad 9\frac{1}{7} \\ \hline \end{array}$

21. $\begin{array}{r} 12 \\ - \quad 4\frac{1}{6} \\ \hline \end{array}$

22. $\begin{array}{r} 2 \\ - \quad 1\frac{2}{3} \\ \hline \end{array}$

23. $\begin{array}{r} 42 \\ - \quad 30\frac{2}{9} \\ \hline \end{array}$

24. $\begin{array}{r} 7 \\ - \quad 5\frac{9}{13} \\ \hline \end{array}$

25. $\begin{array}{r} 19 \\ - \quad 13\frac{3}{8} \\ \hline \end{array}$

26. $\begin{array}{r} 14 \\ - \quad 10\frac{5}{9} \\ \hline \end{array}$

27. $\begin{array}{r} 16 \\ - \quad 8\frac{1}{4} \\ \hline \end{array}$

28. $\begin{array}{r} 15 \\ - \quad 3\frac{5}{7} \\ \hline \end{array}$

1.10   Subtracting Mixed Numbers with Borrowing

Example 12:   Subtract $7\frac{1}{4} - 5\frac{5}{6}$

Step 1:   Rewrite the problem and find a common denominator.

$$7\frac{1}{4} \quad \frac{\times 3}{\times 3} \quad \rightarrow \quad 7\frac{3}{12}$$
$$-5\frac{5}{6} \quad \frac{\times 2}{\times 2} \quad \rightarrow \quad -5\frac{10}{12}$$

Step 2:   You cannot subtract 10 from 3. You must borrow 1 from the 7. The 1 will be in the fraction form $\frac{12}{12}$ which you must add to the $\frac{3}{12}$ you already have, making $\frac{15}{12}$. Subtract whole numbers and simplify.

$$
\begin{array}{r}
\overset{6}{}\overset{15}{\phantom{\frac{3}{12}}} \\
7\frac{3}{12} \\
-5\frac{10}{12} \\
\hline
1\frac{5}{12}
\end{array}
$$

Subtract and simplify.

1. $4\frac{1}{3}$
    $-1\frac{5}{9}$

2. $3\frac{4}{9}$
    $-2\frac{5}{6}$

3. $8\frac{4}{7}$
    $-5\frac{1}{3}$

4. $5\frac{2}{5}$
    $-3\frac{1}{2}$

5. $8\frac{2}{5}$
    $-5\frac{3}{10}$

6. $9\frac{2}{5}$
    $-4\frac{3}{4}$

7. $9\frac{3}{4}$
    $-2\frac{1}{3}$

8. $5\frac{1}{7}$
    $-\frac{2}{3}$

9. $6\frac{1}{5}$
    $-3\frac{3}{8}$

10. $6\frac{5}{6}$
    $-3\frac{4}{5}$

11. $2\frac{2}{9}$
    $-1\frac{3}{4}$

12. $4\frac{7}{10}$
    $-3\frac{1}{3}$

13. $7\frac{3}{5}$
    $-4\frac{5}{6}$

14. $9\frac{3}{8}$
    $-5\frac{1}{2}$

15. $8\frac{1}{9}$
    $-5\frac{1}{3}$

16. $5\frac{1}{6}$
    $-1\frac{2}{3}$

17. $6\frac{5}{6}$
    $-3\frac{1}{3}$

18. $7\frac{2}{3}$
    $-3\frac{5}{6}$

19. $8\frac{4}{7}$
    $-4\frac{3}{4}$

20. $9\frac{3}{4}$
    $-1\frac{1}{5}$

1.11 Multiplying Fractions

Example 13: Multiply $4\frac{3}{8} \times \frac{8}{10}$

Step 1: Change the mixed numbers in the problem to improper fractions. To change $4\frac{3}{8}$ to a mixed number, multiply the denominator by the whole number, add the numerator to this result, and put this total over the old denominator. The denominator is 8, the whole number is 4, and the numerator is 3. To change $4\frac{3}{8}$ to a mixed number, multiply 8×4, add 3, and put this total over 8. The mixed number is then $\frac{35}{8}$. The problem is now to multiply $\frac{35}{8} \times \frac{8}{10}$.

Step 2: When multiplying fractions, you can cancel and simplify terms that have a common factor. The 8 in the first fraction will cancel with the 8 in the second fraction.

$$\frac{35}{\cancel{8}} \times \frac{\cancel{8}}{10}$$

The terms 35 and 10 are both divisible by 5, so

35 simplifies to 7 and 10 simplifies to 2. $\quad \dfrac{\overset{7}{\cancel{35}}}{1} \times \dfrac{1}{\underset{2}{\cancel{10}}}$

Step 3: Multiply the simplified fractions. $\quad \dfrac{7}{1} \times \dfrac{1}{2} = \dfrac{7}{2}$

Step 4: You cannot leave an improper fraction as the answer, so to change $\frac{7}{2}$ back to a mixed number, divide 7 by 2, and put the remainder over the denominator as a fraction. The whole number will be $7 \div 2$ or 3, and since the remainder is 1, the fraction will be $\frac{1}{2}$. The improper fraction $\frac{7}{2}$ is equal to $3\frac{1}{2}$.

Multiply and reduce your answers to lowest terms.

1. $3\frac{1}{3} \times 1\frac{1}{2}$

2. $\frac{3}{8} \times 3\frac{3}{7}$

3. $4\frac{1}{3} \times 2\frac{1}{4}$

4. $4\frac{2}{3} \times 3\frac{3}{4}$

5. $1\frac{1}{2} \times 1\frac{2}{5}$

6. $3\frac{3}{7} \times \frac{5}{6}$

7. $3 \times 6\frac{1}{3}$

8. $1\frac{1}{6} \times 8$

9. $6\frac{2}{5} \times 5$

10. $6 \times 1\frac{3}{8}$

11. $\frac{5}{7} \times 2\frac{1}{3}$

12. $1\frac{2}{5} \times 1\frac{1}{4}$

13. $2\frac{1}{2} \times 5\frac{4}{5}$

14. $7\frac{2}{3} \times \frac{3}{4}$

15. $2 \times 3\frac{1}{4}$

16. $3\frac{1}{8} \times 1\frac{3}{5}$

1.12   Dividing Fractions

Example 14: $1\frac{3}{4} \div 2\frac{5}{8}$

Step 1:   Change the mixed numbers in the problem to improper fractions.
$$1\frac{3}{4} = \frac{(4 \times 1) + 3}{4} = \frac{7}{4} \text{ and } 2\frac{5}{8} = \frac{(8 \times 2) + 5}{8} = \frac{21}{8}.$$
The problem is now $\frac{7}{4} \div \frac{21}{8}$.

Step 2:   Invert (turn upside down) the second fraction and multiply. $\frac{7}{4} \times \frac{8}{21}$

Step 3:   Cancel where possible and multiply. $\frac{1}{1}\frac{7}{4} \times \frac{8}{21}\frac{2}{3} = \frac{2}{3}$

Divide and reduce answers to lowest terms.

1. $2\frac{2}{3} \div 1\frac{7}{9}$

2. $5 \div 1\frac{1}{2}$

3. $1\frac{5}{8} \div 2\frac{1}{4}$

4. $8\frac{2}{3} \div 2\frac{1}{6}$

5. $2\frac{4}{5} \div 2\frac{1}{5}$

6. $3\frac{2}{3} \div 1\frac{1}{6}$

7. $10 \div \frac{4}{5}$

8. $6\frac{1}{2} \div 1\frac{1}{2}$

9. $\frac{3}{5} \div 2$

10. $4\frac{5}{6} \div 1\frac{2}{3}$

11. $9 \div 3\frac{1}{4}$

12. $5\frac{1}{3} \div 2\frac{2}{5}$

13. $4\frac{1}{5} \div \frac{9}{10}$

14. $2\frac{2}{3} \div 4\frac{4}{5}$

15. $3\frac{3}{8} \div 2\frac{5}{7}$

16. $5\frac{1}{4} \div \frac{3}{4}$

1.13   Fraction Word Problems

Solve and reduce answers to lowest terms.

1. Sara buys candy by the pound during the summer. During the first week of summer she buys $1\frac{1}{3}$ pounds of candy, during the second she buys $\frac{3}{4}$ of a pound, and during the third she buys $\frac{4}{5}$ pound. How many pounds did she buy during the first three weeks of summer?

2. Beth has a bread machine that makes a loaf of bread that weighs $1\frac{1}{2}$ pounds. If she makes a loaf of bread for each of her three sisters, how many pounds of bread will she make?

3. Rick chews on a piece of bubble gum for 120 minutes. About every $1\frac{1}{4}$ minutes, each blows a bubble. How many bubbles did Rick make?

4. Juan was competing in a 1000 meter race, but he had to pull out of the race after running $\frac{3}{4}$ of it. How many meters did he run?

5. Tad needs to measure where the free throw line should be in front of his basketball goal. He knows his feet are $1\frac{1}{8}$ feet long and the free-throw line should be 15 feet from the backboard. How many toe-to-toe steps does Tad need to take to mark off 15 feet?

6. Mary gives her puppy a bath and uses $5\frac{1}{2}$ gallons of water. She throws away $3\frac{2}{3}$ gallons of the water. How much water does she have left?

1.14 Comparing the Relative Magnitude of Fractions

Comparing the relative magnitude of fractions using the greater than (>), less than (<), and equal to (=) signs.

Example 15: Compare $\dfrac{3}{4}$ and $\dfrac{5}{8}$

Step 1: Find the lowest common denominator. The lowest common denominator is 8.

Step 2: Change fourths to eighths by multiplying three fourths by two halves, $\dfrac{2 \times 3}{2 \times 4} = \dfrac{6}{8}$.

Step 3: $\dfrac{6}{8} > \dfrac{5}{8}$

Example 16: Compare the mixed numbers $1\dfrac{3}{5}$ and $1\dfrac{2}{3}$.

Step 1: Change the mixed numbers to improper fractions (explained in the previous lesson). $1\dfrac{3}{5} = \dfrac{8}{5}$ and $1\dfrac{2}{3} = \dfrac{5}{3}$

Step 2: Find the lowest common denominator foe the improper fractions. The lowest common denominator is 15.

Step 3: Change fifths to fifteenths and thirds to fifteenths, $\dfrac{3 \times 8}{3 \times 5} = \dfrac{24}{15}$ and $\dfrac{5 \times 5}{5 \times 3} = \dfrac{25}{15}$.

Step 4: $\dfrac{24}{15} < \dfrac{25}{15}$ therefore $1\dfrac{3}{5} < 1\dfrac{2}{3}$.

Fill in the box with the correct sign (>, <, or =).

1. $\dfrac{7}{9}$ ☐ $\dfrac{7}{8}$

2. $\dfrac{6}{7}$ ☐ $\dfrac{5}{6}$

3. $\dfrac{4}{6}$ ☐ $\dfrac{5}{7}$

4. $\dfrac{3}{10}$ ☐ $\dfrac{4}{13}$

5. $\dfrac{5}{8}$ ☐ $\dfrac{4}{11}$

6. $\dfrac{5}{8}$ ☐ $\dfrac{4}{7}$

7. $\dfrac{9}{10}$ ☐ $\dfrac{8}{13}$

8. $\dfrac{2}{13}$ ☐ $\dfrac{1}{10}$

9. $\dfrac{4}{9}$ ☐ $\dfrac{3}{5}$

10. $\dfrac{2}{6}$ ☐ $\dfrac{4}{5}$

11. $\dfrac{7}{12}$ ☐ $\dfrac{6}{11}$

12. $\dfrac{3}{11}$ ☐ $\dfrac{5}{12}$

Chapter 1 Review

Simplify.

1. $\frac{15}{6}$

2. $\frac{24}{5}$

3. $\frac{20}{15}$

4. $\frac{14}{3}$

Reduce.

5. $\frac{9}{27}$

6. $\frac{4}{16}$

7. $\frac{8}{12}$

8. $\frac{12}{18}$

Change to an improper fraction.

9. $5\frac{1}{10}$

10. 7

11. $3\frac{3}{5}$

12. $6\frac{2}{3}$

Add and simplify.

13. $\frac{5}{9} + \frac{7}{9}$

14. $7\frac{1}{2} + 3\frac{3}{8}$

15. $4\frac{4}{15} + \frac{1}{5}$

16. $\frac{1}{7} + \frac{3}{5}$

Subtract and simplify.

17. $10 - 5\frac{1}{8}$

18. $3\frac{1}{3} - \frac{3}{4}$

19. $9\frac{3}{4} - 2\frac{3}{8}$

20. $6\frac{1}{5} - 1\frac{3}{10}$

Multiply and simplify.

21. $1\frac{1}{3} \times 3\frac{1}{2}$

22. $5\frac{3}{5} \times \frac{7}{6}$

23. $4\frac{4}{6} \times 1\frac{5}{7}$

24. $\frac{2}{3} \times \frac{5}{6}$

Divide and simplify.

25. $\frac{1}{2} \div \frac{4}{5}$

26. $6\frac{6}{7} \div 2\frac{2}{3}$

27. $3\frac{5}{6} \div 11\frac{1}{2}$

28. $1\frac{1}{3} \div 3\frac{1}{5}$

Find the greatest common factor for the following sets of numbers.

29. 9 and 15

30. 12 and 16

31. 10 and 25

32. 8 and 24

Find the least common multiple for the following sets of numbers.

33. 8 and 12

34. 5 and 9

35. 4 and 10

36. 6 and 8

Find the missing numerators.

37. $\dfrac{4}{11} = \dfrac{}{55}$

39. $\dfrac{2}{7} = \dfrac{}{35}$

41. $\dfrac{3}{5} = \dfrac{}{45}$

43. $\dfrac{4}{9} = \dfrac{}{54}$

38. $\dfrac{3}{8} = \dfrac{}{48}$

40. $\dfrac{7}{10} = \dfrac{}{80}$

42. $\dfrac{5}{12} = \dfrac{}{48}$

44. $\dfrac{3}{11} = \dfrac{}{121}$

45. The Vargas family is hiking a $23\frac{1}{3}$ mile trail. The first day, they hiked $10\frac{1}{2}$ miles. How much further do they have to go to complete the trail?

46. Jena walks $\frac{1}{5}$ of a mile to a friend's house, $1\frac{1}{3}$ miles to the store, and $\frac{3}{4}$ of a mile back home. How far does Jena walk?

47. Cory uses $2\frac{4}{5}$ gallons of paint to mark one mile of this year's spring road race. How many gallons will he use to mark the entire $6\frac{1}{4}$ mile course?

Chapter 2
Decimals

This chapter covers the following Georgia Performance Standards:

M6N	Number and Operations	M6N1.f, g
M6P	Process Skills	M6P1.a, b
		M6P3.a, c, d
		M6P4.c
		M6P5.a, b, c

2.1   Adding Decimals

Example 1:   Find $0.9 + 2.5 + 63.17$

Step 1:   When you add decimals, first arrange the numbers in columns with the decimal points under each other.

$$\begin{array}{r} 0.9 \\ 2.5 \\ 63.17 \end{array}$$

Step 2:   Add 0's here to keep your columns straight. ⟶

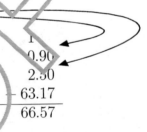

Step 3:   Start at the right and add each column. Remember to carry when necessary. Bring down the decimal point.

$$\begin{array}{r} 1 \\ 0.90 \\ 2.50 \\ +\ 63.17 \\ \hline 66.57 \end{array}$$

Add. Be sure to write the decimal point in your answer.

1. $5.3 + 6.02 + 0.73$

2. $0.235 + 6.2 + 3.27$

3. $7.542 + 10.5 + 4.57$

4. $\$5.87 + \$7.52$

5. $\$4.68 + \$9.47$

6. $5.08 + 11.2 + 6.075$

7. $5.14 + 2.3 + 5.097$

8. $4.9 + 15.71 + 0.254$

9. $\$3.75 + \$18.90$

10. $\$64.95 + \$4.65$

11. $1.25 + 4.1 + 10.007$

12. $15.4 + 5.074 + 3.15$

13. $45.23 + 9.5 + 0.693$

14. $\$8.63 + \$12.50$

15. $\$6.87 + \$27.23$

16. $0.23 + 5.9 + 12$

17. $8.5784 + 10.03$

18. $85.7 + 205.952$

19. $\$98.45 + \$8.89$

20. $\$7.77 + \$11.19$

21. $3.27 + 8.0054 + 1.1$

2.2   Subtracting Decimals

Example 2:   Find $14.9 - 0.007$

Step 1: When you subtract decimals, arrange the numbers in columns with the decimal points under each other.

$$\begin{array}{r} 14.9 \\ -\ 0.007 \\ \hline \end{array}$$

Step 2: You must fill in the empty places with 0's so that both numbers have the same number of digits after the decimal point.

$$\begin{array}{r} 14.900 \\ -\ 0.007 \\ \hline \end{array}$$

Step 3: Start at the right and subtract each column. Remember to borrow when necessary.

$$\begin{array}{r} 89_1 \\ 14.9\cancel{0}0 \\ -\ 0.007 \\ \hline 14.893 \end{array}$$

Subtract. Be sure to write the decimal point in your answer.

1. $5.25 - 4.7$

2. $23.657 - 9.85$

3. $\$56.54 - \$17.02$

4. $\$294.78 - \$80.99$

5. $\$76.00 - \$68.99$

6. $58.6 - 9.153$

7. $405.97 - 7.325$

8. $\$40.09 - \$9.99$

9. $\$115.45 - \$4.79$

10. $\$45.18 - \$23.65$

11. $12.96 - 7.32$

12. $19.2 - 8.63$

13. $8.123 - 5.096$

14. $\$14.32 - \$0.58$

15. $\$30.00 - \$22.95$

16. $15.780 - 6.32$

17. $478.63 - 99.2$

18. $\$15.45 - \$8.58$

19. $102.5 - 1.079$

20. $7.054 - 3.009$

21. $12.42 - 3.235$

2.3   Multiplication of Decimals

Example 3: 56.2×0.17

Step 1:   Set up the problem as if you were multiplying whole numbers.

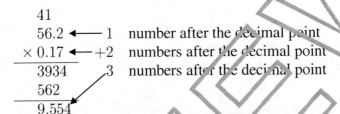

$$\begin{array}{r} 56.2 \\ \times\, 0.17 \\ \hline \end{array}$$

Step 2:   Multiply as if you were multiplying whole numbers.

$$\begin{array}{r} 41 \\ 56.2 \longleftarrow 1 \text{ number after the decimal point} \\ \times\, 0.17 \longleftarrow +2 \text{ numbers after the decimal point} \\ \hline 3934 \quad 3 \text{ numbers after the decimal point} \\ 562 \\ \hline 9.554 \end{array}$$

Step 3:   Count how many numbers are after the decimal points in the problem. In this problem, 2, 1, and 7 come after decimal points, so the answer must also have three numbers after the decimal point.

Multiply.

1. 15.2×3.5
2. 9.54×5.3
3. 5.72×6.3
4. 4.8×3.2

5. 15.8×2.2
6. 4.5×7.1
7. 0.052×0.33
8. 4.12×6.8

9. 23.65×9.2
10. 1.54×0.43
11. 0.47×6.1
12. 1.3×1.57

13. 16.4×0.5
14. 0.87×3.21
15. 5.94×0.65
16. 7.8×0.23

2.4   Division of Decimals by Whole Numbers

Example 4: $52.26 \div 6$

Step 1:   Copy the problem as you would for whole numbers. Copy the decimal point directly above in the place for the answer.

$$6\,\overline{)\,52.26}$$

Step 2:   Divide the same way as you would with whole numbers.

$$\begin{array}{r} 8\,.\,71 \\ 6\,\overline{)\,52\,.\,26} \\ 48 \\ \hline 4\ \ 2 \\ -\ \ 4\ \ 2 \\ \hline 6 \\ -\ \ 6 \\ \hline 0 \end{array}$$

Divide. Remember to copy the decimal point directly about the place for the answer.

1. $42.75 \div 3$ 5. $12.50 \div 2$ 9. $72.36 \div 4$ 13. $102.5 \div 5$

2. $74.16 \div 6$ 6. $224.64 \div 52$ 10. $379.5 \div 15$ 14. $113.4 \div 9$

3. $81.50 \div 25$ 7. $183.04 \div 52$ 11. $152.25 \div 21$ 15. $585.14 \div 34$

4. $82.46 \div 14$ 8. $281.52 \div 23$ 12. $40.375 \div 19$ 16. $93.6 \div 24$

2.5 Division of Decimals by Decimals

Example 5: $374.5 \div 0.07$

Step 1:   Copy the problem as you would for whole numbers.

$$0.07\overline{)374.5}$$

(Divisor points to 0.07, Dividend points to 374.5)

Step 2:   You cannot divide by a decimal number. You must move the decimal point in the divisor 2 places to the right to make it a whole number. The decimal point in the dividend must also move to the right the same number of places. Notice that in this example, you must add a 0 to the dividend.

$$0.07.\overline{)374.50.}$$

Step 3:   The problem now becomes $37450 \div 7$. Copy the decimal point from the dividend straight above in the place for the answer.

```
        5 3 5 0 .
    7 ) 3 7 4 5 0 .
      - 3 5
        ----
          2 4
        - 2 1
          ----
            3 5
          - 3 5
            ----
              0 0
```

Divide. Remember to move the decimal points.

1. $0.676 \div 0.013$ 5. $18.46 \div 1.3$ 9. $154.08 \div 1.8$ 13. $4.8 \div 0.08$

2. $70.32 \div 0.08$ 6. $14.6 \div 0.002$ 10. $0.4374 \div 0.003$ 14. $1.2 \div 0.024$

3. $\$54.60 \div 0.84$ 7. $\$125.25 \div 0.75$ 11. $292.9 \div 0.29$ 15. $15.725 \div 3.7$

4. $\$10.35 \div 0.45$ 8. $\$33.00 \div 1.65$ 12. $6.375 \div 0.3$ 16. $\$167.50 \div 0.25$

2.6   Changing Fractions to Decimals

Example 6:   Change $\frac{1}{8}$ to a decimal.

Step 1:   To change a fraction to a decimal, simply divide the top number by the bottom number.

$$8\,\overline{)1}$$

Step 2:   Add a decimal point and a 0 after the 1 and divide.

$$\begin{array}{r} 0.1 \\ 8\,\overline{)1.0} \\ -8 \\ \hline 2 \end{array}$$

Step 3:   Continue adding 0's and dividing until there is no remainder.

$$\begin{array}{r} 0.125 \\ 8\,\overline{)1.000} \\ -8 \\ \hline 20 \\ -16 \\ \hline 40 \\ -40 \\ \hline 0 \end{array}$$

In some problems the number after the decimal point begins to repeat. Take, for example, the fraction $\frac{4}{11}$, $4 \div 11 = 0.363636$ and the 36 keeps repeating forever. To show that 36 repeats, simply write a bar above the numbers that repeat, $0.\overline{36}$.

Change the following fractions to decimals.

1. $\frac{4}{5}$ 5. $\frac{1}{10}$ 9. $\frac{3}{5}$ 13. $\frac{7}{9}$ 17. $\frac{3}{16}$

2. $\frac{2}{3}$ 6. $\frac{5}{8}$ 10. $\frac{7}{10}$ 14. $\frac{9}{10}$ 18. $\frac{3}{4}$

3. $\frac{1}{2}$ 7. $\frac{5}{6}$ 11. $\frac{4}{11}$ 15. $\frac{1}{4}$ 19. $\frac{8}{9}$

4. $\frac{5}{9}$ 8. $\frac{1}{6}$ 12. $\frac{1}{9}$ 16. $\frac{3}{8}$ 20. $\frac{5}{12}$

2.7   Changing Mixed Numbers to Decimals

If there is a whole number with a fraction, write the whole number to the left of the decimal point. Then change the fraction to a decimal.

Examples: $4\dfrac{1}{10} = 4.1$ $16\dfrac{2}{3} = 16.\overline{6}$ $12\dfrac{7}{8} = 12.875$

Change the following mixed numbers to decimals.

1. $5\dfrac{2}{3}$ 5. $30\dfrac{1}{3}$ 9. $6\dfrac{4}{5}$ 13. $7\dfrac{1}{4}$ 17. $10\dfrac{1}{10}$

2. $8\dfrac{5}{11}$ 6. $3\dfrac{1}{2}$ 10. $13\dfrac{1}{2}$ 14. $12\dfrac{1}{3}$ 18. $20\dfrac{2}{5}$

3. $15\dfrac{3}{5}$ 7. $1\dfrac{7}{8}$ 11. $12\dfrac{4}{5}$ 15. $1\dfrac{5}{8}$ 19. $4\dfrac{9}{10}$

4. $13\dfrac{2}{3}$ 8. $4\dfrac{9}{100}$ 12. $11\dfrac{5}{8}$ 16. $2\dfrac{3}{4}$ 20. $5\dfrac{4}{11}$

2.8   Changing Decimals to Fractions

Example 7:   Change 0.25 to a fraction.

Step 1:   Copy the decimal without the point. This will be the top number of the fraction. $\dfrac{25}{}$

Step 2:   The bottom number is a 1 with as many 0's after it as there are digits in the top number. $\dfrac{25 \leftarrow \text{Two digits}}{100 \leftarrow \text{Two 0's}}$

Step 3:   You then need to reduce the fraction. $\dfrac{25}{100} = \dfrac{1}{4}$

Examples: $0.2 = \dfrac{2}{10} = \dfrac{1}{5}$ $0.65 = \dfrac{65}{100} = \dfrac{13}{20}$ $0.125 = \dfrac{125}{1000} = \dfrac{1}{8}$

Change the following decimals to fractions.

1. 0.55   5. 0.75   9. 0.71   13. 0.35

2. 0.6   6. 0.82   10. 0.42   14. 0.96

3. 0.12   7. 0.3   11. 0.56   15. 0.125

4. 0.9   8. 0.42   12. 0.24   16. 0.375

2.9   Changing Decimals with Whole Numbers to Mixed Numbers

Example 8:   Change 14.28 to a mixed number.

Step 1:   Copy the portion of the number that is whole.   14

Step 2:   Change .28 to a fraction. $14\frac{28}{100}$

Step 3:   Reduce the fraction. $14\frac{28}{100} = 14\frac{7}{25}$

Change the following decimals to mixed numbers.

1. 7.125
2. 99.5
3. 2.13
4. 5.1

5. 16.95
6. 3.625
7. 4.42
8. 15.84

9. 6.7
10. 15.425
11. 15.8
12. 8.16

13. 13.9
14. 32.65
15. 17.25
16. 9.82

2.10   Decimal Word Problems

1. Micah can have his bike fixed for $19.99, or he can buy the new part for his bike and replace it himself for $8.79. How much would he save by fixing his bike himself?

2. Megan buys 5 boxes of cookies for $3.75 each. How much does she spend?

3. Will subscribes to a monthly sports magazine. His one-year subscription costs $29.97. If he pays for the subscription in 3 equal installments, how much is each payment?

4. Pat purchases 2.5 pounds of jelly beans at $0.98 per pound. What is the total cost of the jelly beans?

5. The White family took $650 cash with them on vacation. At the end of their vacation, they had $4.67 left. How much cash did they spend on vacation?

6. Ace Middle School spends $1443.20 on 55 math books. How much does each book cost?

7. The Junior Beta Club needs to raise $1518.75 to go to a national convention. If they decide to sell candy bars at $1.25 each, how many must they sell to meet their goal?

8. Fleta owns a candy store. On Monday, she sold 6.5 pounds of chocolate, 8.34 pounds of jelly beans, 4.9 pounds of sour snaps, and 5.64 pounds of yogurt-covered raisins. How many pounds of candy did she sell in total?

9. Randal purchased a rare coin collection for $1803.95. He sold it at auction for $2700. How much money did he make on the coins?

10. A leather jacket that normally sells for $259.99 is on sale now for $197.88. How much can you save if you buy it now?

Chapter 2 Review

Add.

1. $12.589 + 5.62 + 0.9$

2. $7.8 + 10.24 + 1.903$

3. $152.64 + 12.3 + 0.024$

Subtract.

4. $18.547 - 9.62$

5. $1.85 - 0.093$

6. $45.2 - 37.9$

Multiply.

7. 4.58×0.025

8. 0.879×1.7

9. 30.7×0.0041

Divide.

10. $17.28 \div 0.054$

11. $174.66 \div 1.23$

12. $2.115 \div 9$

Change to a fraction.

13. 0.55

14. 0.84

15. 0.32

Change to a mixed number.

16. 7.375

17. 9.6

18. 12.25

Change to a decimal.

19. $5\frac{2}{25}$

20. $\frac{7}{100}$

21. $10\frac{2}{3}$

Use $>$ and $<$ to compare the following.

22. $0.123 \square 0.21234$

23. $0.025 \square 0.125$

24. Gene works for his father sanding wooden rocking chairs. He earns \$6.35 per chair. How many chairs does he need to sand in order to buy a portable radio/CD player for \$146.05?

25. Margo's Mint Shop has a machine that produces 4.35 pounds of mints per hour. How many pounds of mints are produced in each 8-hour shift?

26. Carter's Junior High track team runs the first leg of a 400-meter relay race in 10.23 seconds, the second leg in 11.4 seconds, the third leg in 10.77 seconds, and the last leg in 9.9 seconds. How long does it take for them to complete the race?

Chapter 3
Percents

This chapter covers the following Georgia Performance Standards:

M6N	Number and Operations	M6N1.f, g
M6P	Process Skills	M6P1.a, b, c
		M6P3.a, c, d
		M6P4.a, b, c
		M6P5.a, b, c

3.1   Changing Percents to Decimals and Decimals to Percents

To change a **percent** to a **decimal**, move the **decimal** point two places to the left, and drop the **percent** sign. If there is no decimal point shown, it is understood to be after the number and before the percent sign. Sometimes you will need to add a "0". (See 5% below.)

Example 1: $14\% = 0.14$ $5\% = 0.05$ $100\% = 1$ $103\% = 1.03$

    ↗
    (decimal point)

Change the following percents to decimal numbers.

1. 18%	4. 65%	7. 2%	10. 55%	13. 66%	16. 25%	19. 50%
2. 23%	5. 4%	8. 119%	11. 80%	14. 13%	17. 410%	20. 99%
3. 9%	6. 45%	9. 7%	12. 17%	15. 5%	18. 1%	21. 107%

To change a decimal to a percent, move the decimal two places to the right, and add a percent sign. You may need to add a "0". (See 0.8 below.)

Example 2: $0.62 = 62\%$ $0.07 = 7\%$ $0.8 = 80\%$ $0.166 = 16.6\%$ $1.54 = 154\%$

Change the following decimal numbers to percents.

22. 0.15	25. 0.22	28. 0.648	31. 0.86	34. 0.48	37. 0.375	40. 0.3
23. 0.62	26. 0.35	29. 0.044	32. 0.29	35. 3.089	38. 5.09	41. 2.9
24. 1.53	27. 0.375	30. 0.58	33. 0.06	36. 0.042	39. 0.75	42. 0.06

3.2   Changing Percents to Fractions and Fractions to Percents

Example 3:   Change 15% to a fraction.

Step 1:   Copy the number without the percent sign. 15 is the top number of the fraction.

Step 2:   The bottom number of the fraction is 100.

$$15\% = \frac{15}{100}$$

Step 3:   Reduce the fraction. $\frac{15}{100} = \frac{3}{20}$

Change the following percents to fractions and reduce.

1. 50%	5. 52%	9. 18%	13. 16%	17. 99%
2. 13%	6. 63%	10. 3%	14. 1%	18. 30%
3. 22%	7. 75%	11. 25%	15. 79%	19. 15%
4. 95%	8. 91%	12. 5%	16. 40%	20. 84%

Example 4:   Change $\frac{7}{8}$ to a percent.

Step 1:   Divide 7 by 8. Add as many 0's as necessary.

$$\begin{array}{r} 0.875 \\ 8\overline{\smash{)}7.000} \\ -\underline{64} \\ 60 \\ -\underline{56} \\ 40 \\ -\underline{40} \\ 0 \end{array}$$

Step 2:   Change the decimal answer, 0.875, to a percent by moving the decimal point 2 places to the right.

$$\frac{7}{8} = 0.875 = 87.5\%$$

Change the following fractions to percents.

1. $\frac{1}{5}$	4. $\frac{3}{8}$	7. $\frac{1}{10}$	10. $\frac{3}{4}$	13. $\frac{1}{16}$	16. $\frac{3}{4}$
2. $\frac{5}{8}$	5. $\frac{3}{16}$	8. $\frac{4}{5}$	11. $\frac{1}{8}$	14. $\frac{1}{4}$	17. $\frac{2}{5}$
3. $\frac{7}{16}$	6. $\frac{19}{100}$	9. $\frac{15}{16}$	12. $\frac{5}{16}$	15. $\frac{4}{100}$	18. $\frac{16}{25}$

3.3   Changing Percents to Mixed Numbers and Mixed Numbers to Percents

Example 5:   Change 218% to a fraction.

Step 1:   Copy the number without the percent sign. 218 is the top number of the fraction.

Step 2:   The bottom number of the fraction is 100.

$$218\% = \frac{218}{100}$$

Step 3:   Reduce the fraction, and convert to a mixed number. $\frac{218}{100} = \frac{109}{50} = 2\frac{9}{50}$

Change the following percents to mixed numbers.

1. 150%	6. 163%	11. 205%	16. 340%
2. 113%	7. 275%	12. 405%	17. 199%
3. 222%	8. 191%	13. 516%	18. 300%
4. 395%	9. 108%	14. 161%	19. 125%
5. 252%	10. 453%	15. 179%	20. 384%

Example 6:   Change $5\frac{3}{8}$ to a percent.

Step 1:   Divide 3 by 8. Add as many 0's as necessary.

$$\begin{array}{r} 0.375 \\ 8\overline{)3.000} \\ -2\ 4 \\ \hline 60 \\ -56 \\ \hline 40 \\ -40 \\ \hline 0 \end{array}$$

Step 2:   So, $5\frac{3}{8} = 5.375$. Change the decimal answer to a percent by moving the decimal point 2 places to the right.

$$5\frac{3}{8} = 5.375 = 537.5\%$$

Change the following mixed and whole numbers to percents.

1. $5\frac{1}{2}$	4. $3\frac{1}{4}$	7. $1\frac{3}{10}$	10. $2\frac{13}{25}$	13. $1\frac{3}{16}$	16. $4\frac{4}{5}$
2. $8\frac{3}{4}$	5. $4\frac{7}{8}$	8. $6\frac{1}{5}$	11. $1\frac{1}{8}$	14. $1\frac{1}{16}$	17. $3\frac{2}{5}$
3. 1	6. 3	9. 4	12. 2	15. 5	18. 6

3.4   Comparing the Relative Magnitude of Numbers

When comparing the relative magnitude of numbers, the greater than ($>$), less than ($<$), and the equal to ($=$) signs are the ones most frequently used. The simplest way to compare numbers that are in different notations, like percent, decimals, and fractions, is to change all of them to one notation. Decimals are the easiest to compare.

Example 7:   Which is larger: $1\frac{1}{4}$ or 1.3?

Step 1:   Change $1\frac{1}{4}$ to a decimal. $\frac{1}{4} = 0.25$, so $1\frac{1}{4} = 1.25$.

Step 2:   Compare the two values in decimal form.
$1.25 < 1.3$, so 1.3 is the larger of the two values.

Example 8:   Which is smaller: 60% or $\frac{2}{3}$?

Step 1:   Change both values to decimals.
$60\% = 0.6$ and $\frac{2}{3} = 0.\overline{66}$

Step 2:   Compare the two values in decimal form.
0.6 is smaller than $0.\overline{66}$, so $60\% < \frac{2}{3}$.

Fill in each box with the correct sign.

1. $23.4\ \square\ 23\frac{1}{2}$

2. $17\%\ \square\ .17$

3. $\frac{3}{8}\ \square\ 37.5\%$

4. $25\%\ \square\ \frac{2}{10}$

5. $234\%\ \square\ 23.4$

6. $\frac{1}{7}\ \square\ 14\%$

7. $12.95\ \square\ 13\frac{8}{9}$

8. $4.0\ \square\ 40\%$

9. $25\%\ \square\ \frac{3}{2}$

10. $\frac{15}{4}\ \square\ 300\%$

11. $6\%\ \square\ \frac{1}{16}$

12. $1.33\ \square\ \frac{4}{3}$

13. $.8\ \square\ \frac{4}{5}$

14. $75\%\ \square\ \frac{3}{4}$

15. $\frac{5}{8}\ \square\ 62\%$

Compare the sums, differences, products, and quotients below. Fill in each box with the correct sign.

16. $(32 + 15)\ \square\ (65 - 17)$

17. $(45 - 13)\ \square\ (31 + 9)$

18. $(24 \div 4)\ \square\ (24 \div 6)$

19. $(48 \div 6)\ \square\ (4 \times 3)$

20. $(4 \times 3)\ \square\ (48 \div 6)$

21. $(13 \times 4)\ \square\ (5 \times 17)$

22. $[(1 + 3) + 5]\ \square\ [5 + (3 + 1)]$

23. $[1 + (3 + 5)]\ \square\ [(5 - 3) + 1]$

24. $(25 \div 5)\ \square\ (5 \times 5)$

25. $(6 + 4 \div 2)\ \square\ [(6 + 4) \div 2]$

3.5   Changing to Percent Word Problems

Example 9:   Three out of four students prefer pizza for lunch over hot dogs. What percent of the students prefer pizza?

Step 1:   Change three out of four to a fraction. $\frac{3}{4}$

Step 2:   Change $\frac{3}{4}$ to a percent. $\frac{3}{4} = 75\%$ of the students prefer pizza.

For each of the following, find the percent.

1. 18 out of 24 students take a music class in middle school.

2. 6 out of the 30 students get a grade of an A in the class.

3. 3 out of 10 students ride the bus to school.

4. 16 out of 20 students in my first period ate breakfast.

5. Out of 12 cats, 3 of them have stripes.

6. In a neighborhood store, 5 out of 50 customers have an infant in their cart.

7. 240 out of 500 people at the fair are over 21.

8. In Alaska, 9 months out of 12 have temperatures below 0° C.

9. He hits the ball 360 times out of 500 times at bat.

10. She scores 552 out of a possible 600 points on her math test.

11. 16 out of 40 pieces in the box of candy are covered in dark chocolate.

12. 680 out of 800 students have not had a cold this year.

13. 7 out of 8 of the flowers are pink.

14. 312 out of 600 men at the baseball game have on earmuffs.

15. 824 out of 1000 tulips are red.

16. Kristen sells 475 out of her 500 boxes of Girl Scout cookies.

17. 459 out of 675 students in the 6th grade went to the spring dance this year.

18. In a bag of 60 candy coated chocolate pieces, 9 are yellow.

3.6   Finding the Percent of the Total

Example 10:   There are 75 customers at Billy's gas station this morning. Thirty-two percent use a credit card to make their purchases. How many customers used credit cards this morning at Billy's?

Step 1:   Change 32% to a decimal.   0.32

Step 2:   Multiply by the total number mentioned.

$$\begin{array}{r} 0.3\,2 \\ \times 7\,5 \\ \hline 16\,0 \\ 224 \\ \hline 24.00 \end{array}$$

24 customers used credit cards.

Answer the following questions.

1. Eighty-five percent of Mrs. Coomer's math class pass her final exam. There are 40 students in her class. How many pass?

2. Fifteen percent of a bag of chocolate candies have a red coating on them. How many red pieces are in a bag of 60 candies?

3. Sixty-eight percent of Valley Creek School students attend this year's homecoming dance. There are 675 students. How many attend the dance?

4. Out of the 4,500 people who attend the rock concert, forty-six percent purchase a T-shirt. How many people buy T-shirts?

5. Nina sells ninety-five percent of her 500 cookies at the bake sale. How many cookies does she sell?

6. Twelve percent of yesterday's customers purchased premium-grade gasoline from GasCo. If GasCo had 200 customers, how many purchased premium-grade gasoline?

7. The Candy Shack sells 138 pounds of candy on Tuesday. Fifty-two percent of the candy is jelly beans. How many pounds of jelly beans are sold Tuesday?

8. A fund-raiser at the school raises $617.50. Ninety-four percent goes to local charities. How much money goes to charities?

9. Out of the company's $6.5 million profit, eight percent will be paid to shareholders. How much will be paid to the shareholders?

10. Ted's Toys sells seventy-five percent of its stock of stuffed bean animals on Saturday. If Ted's Toys had 620 originally in stock, how many are sold on Saturday?

3.7   Finding the Percent Increase or Decrease

Example 11:   Office Supply Co. purchases paper wholesale for $18.00 per case. They sell the paper for $20.00 per case. By what percent does the store increase the price of the paper (or what is the percent markup)?

$$\text{Percent Change} = \frac{\text{Amount of Change}}{\text{Original Amount}}$$

Step 1:   Find the amount of change. In this problem, the price was marked up $2.00. The amount of change is 2.

Step 2:   Divide the amount of change, 2, by the wholesale cost, 18. $\frac{2}{18} = 0.111$

Step 3:   Change the decimal, 0.111, to a percent. $0.111 = 11.1\%$

Example 12:   The price of gas goes from $2.40 per gallon to $1.30. What is the percent of decrease in the price of gas?

$$\text{Percent Change} = \frac{\text{Amount of Change}}{\text{Original Amount}}$$

Step 1:   Find the amount of change. In this problem, the price is decreased $1.10. The amount of change is $1.10.

Step 2:   Divide the amount of change, 1.10, by the original cost, 2.40. $\frac{1.10}{2.40} = 0.46$

Step 3:   Change the decimal, 0.46, to a percent. $0.46 = 46\%$, the price of gas has decreased 46%.

Find the percent increase or decrease to the nearest percent for each of the problems below.

1. Mary was making $25,000 per year. Her boss gives her a $3,000 raise. What percent increase is that?

2. Last week Matt's total sales were $12,000. This week his total sales were only $2,000. By what percent did his sales for this week decrease?

3. Eric cuts lawns for $16.00. Next year, he will charge a $2.00 more per lawn. What percent increase will he charge?

4. At an office supply store, pens *are* marked down from $1.50 to $1.20. What percent discount is that?

5. Cowboys buy boots wholesale for $103.35. They sell the boots in their store for $159. What percent is the markup on the boots?

6. Blakeville has a population of 1600. According to the last census, Blakeville had a population of 1850. What has been the percent decrease in population?

7. Last year, Roswell Elementary School had 680 graduates. This year they graduated 812. What has been the percent increase in graduates?

8. Michi's father gets a new job that pays $52,000 per year. That is $16,000 more than his last job. What percent pay increase is that?

3.8   Tips and Commissions

Vocabulary

Tip: A **tip** is money given to someone doing a service for you such as a server, hair stylist, porter, cab driver, grocery bagger, etc.

Commission: In many businesses, sales people are paid on **commission** - a percent of the total sales they make.

Problems requiring you to figure a tip, commission, or percent of a total are all done in the same way.

Example 13:   Ramon makes a 4% commission on an $8,000 pickup truck he sold. How much is his commission?

$$
\begin{array}{rr}
\text{TOTAL COST} & \$8,000 \\
\times \quad \text{RATE OF COMMISSION} \quad \times & 0.04 \\
\hline
\text{COMMISSION} & \$320.00
\end{array}
$$

Solve each of the following problems.

1. Mia makes 12% commission on all her sales. This month she sells $9,000 worth of merchandise. What is her commission?

2. Marcus gives 25% of his income to his parents to help cover expenses. He earns $340 per week. How much money does he give his parents?

3. Jan pays $640 per month for rent. If rent goes up by 5%, how much can Jan expect to pay monthly next year?

4. The total bill at Jake's Catfish Place comes to $35.80. Palo wants to leave a 15% tip. How much money will he have to leave for the tip?

5. Rami makes $2,400 per month and puts 6% in a savings account. How much does he save per month?

6. Christina makes $2,550 per month. Her boss promises her a 7% raise. How much more will she make per month?

7. Out of 150 math students, 86% pass. How many students pass math class?

8. Marta sells Sue Anne Cosmetics and gets 20% commission on all her sales. Last month, she sold $560.00 worth of cosmetics. How much was her commission?

3.9   Finding the Amount of a Discount

Sale prices are sometimes marked 30% off, or better yet 50% off. A 30% **discount** means you will pay 30% less than the original price. How much money you will save is also known as the amount of the discount. Read the example below to learn to figure the amount of a discount.

Example 14:   A $179.00 chair is on sale for 30% off. How much can I save if I buy it now?

Step 1:   Change 30% to a decimal. $30\% = 0.30$

Step 2:   Multiply the original price by the discount.

$$
\begin{array}{r}
\textbf{ORIGINAL PRICE} \quad \$179.00 \\
\times \quad \underline{\textbf{\% DISCOUNT}} \quad \times \quad \underline{0.30} \\
\textbf{SAVINGS} \quad \$53.70
\end{array}
$$

Practice finding the amount of discount. Round off answers to the nearest penny.

1. Tubby Telephones is offering a 25% discount on phones purchased on Tuesday. How much can you save if you buy a phone on Tuesday regularly priced at $225.00 any other day of the week?

2. The regular price for a garden rake is $10.97 at Sly's Super Store. This week, Sly is offering a 30% discount. How much is the discount on the rake?

3. Christine buys a sweater regularly priced at $26.80 with a coupon for 20% off any sweater. How much does she save?

4. The software that Myoshi needs for her computer is priced at $69.85. If she waits until a store offers it for 20% off, how much will she save?

5. Ty purchases jeans that were priced at $23.97. He received a 15% employee discount. How much does he save?

6. The Bakery Company offers a 60% discount on all bread made the day before. How much can you save on a $2.40 loaf made today if you wait until tomorrow to buy it?

7. A furniture store advertises a 40% off sale on all items. How much would the discount be on a $2530 bedroom set?

8. Sharta buys a $4.00 bottle of nail polish on sale for 30% off. What is the dollar amount of the discount?

9. How much is the discount on a $350 racing bike marked 15% off?

10. Raymond receives a 2% discount from his credit card company on all purchases made with the credit card. What is his discount on $1575.50 worth of purchases?

3.10   Finding the Discounted Sale Price

To find the discounted sale price, you must go one step further than shown on the previous page. Read the example below to learn how to figure **discount** prices.

Example 15:   A $74.00 chair is on sale for 25% off. How much will it cost if I buy it now?

Step 1:   Change 25% to a decimal.

25% = 0.25

Step 2:   Multiply the original price by the discount.

$$
\begin{array}{rr}
\textbf{ORIGINAL PRICE} & \$74.00 \\
\times \quad \textbf{\% DISCOUNT} & \times \quad 0.25 \\
\hline
\textbf{SAVINGS} & \$18.50 \\
\end{array}
$$

Step 3:   Subtract the savings amount from the original price to find the sale price.

$$
\begin{array}{rr}
\textbf{ORIGINAL PRICE} & \$74.00 \\
- \quad \textbf{SAVINGS} & - \quad 18.50 \\
\hline
\textbf{SALE PRICE} & \$55.50 \\
\end{array}
$$

Figure the sale price of the items below. The first one is done for you.

ITEM	PRICE	%OFF	MULTIPLY	SUBTRACT	SALE PRICE
1. pen	$1.50	20%	$1.50 \times .2 = \$0.30$	$1.50 - 0.30 = 1.20$	$1.20
2. recliner	$325	25%			
3. juicer	$55	15%			
4. blanket	$14	10%			
5. earrings	$2.40	20%			
6. figurine	$8	15%			
7. boots	$159	35%			
8. calculator	$80	30%			
9. candle	$6.20	50%			
10. camera	$445	20%			
11. VCR	$235	25%			
12. video game	$25	10%			

3.11   Sales Tax

Example 16:   The total price of a sofa is $560.00 × 6% **sales tax**. How much is the sales tax? What is the total cost?

Step 1:   You will need to change 6% to a decimal.

$6\% = 0.06$

Step 2:   Simply multiply the cost, $560, by the tax rate, 6%. $560 \times 0.06 = 33.6$. The answer will be $33.60. (You need to add a 0 to the answer. When dealing with money, there must be two places after the decimal point.)

	COST	$560
×	6% TAX	× 0.06
	SALES TAX	$33.60

Step 3:   Add the sales tax amount, $33.60, to the cost of the item sold, $560. This is the total cost.

	COST	$560.00
+	SALES TAX	+ 33.60
	TOTAL COST	$593.60

Note: When the answer to the question involves money, you always need to round off the answer to the nearest hundredth (2 places after the decimal point). Sometimes you will need to add a zero.

Figure the total costs in the problems below. The first one is done for you.

	ITEM	PRICE	% TAX	MULTIPLY	PRICE PLUS TAX	TOTAL
1.	jeans	$42	7%	$42 × 0.07 = $2.94	42 + 2.94 = 44.94	$44.94
2.	truck	$17,495	6%			
3.	film	$5.89	8%			
4.	T-shirt	$12	5%			
5.	football	$36.40	4%			
6.	soda	$1.78	5%			
7.	4 tires	$105.80	10%			
8.	clock	$18	6%			
9.	burger	$2.34	5%			
10.	software	$89.95	8%			

Chapter 3 Review

Change the following percents to decimals.

1. 45%
2. 219%
3. 22%
4. 1.25%

Change the following decimals to percents.

5. 0.52
6. 0.64
7. 1.09
8. 0.625

9. What is 1.65 written as a percent?

10. Change 5.65 to a percent.

Change the following percents to fractions.

11. 25%
12. 3%
13. 68%
14. 102%

Change the following fractions to percents.

15. $\frac{9}{10}$
16. $\frac{5}{16}$
17. $\frac{1}{8}$
18. $\frac{1}{4}$

Use the >, <, and = signs to make the following correct.

19. $\frac{5}{6}$ ☐ $\frac{4}{5}$
20. $\frac{3}{7}$ ☐ $\frac{1}{8}$
21. $\frac{4}{15}$ ☐ $\frac{5}{16}$
22. $\frac{3}{4}$ ☐ $\frac{13}{16}$

Fill in the box with the correct sign <, >, or =.

23. $(54 \div 6)$ ☐ (8×7)

24. $[3 + (5 - 2)]$ ☐ $[1 + (6 \div 2)]$

25. The school recieves 56% of the total sale at the end of a fundraiser. If a fundraiser makes $564,000, how much money does the school receive?

26. Celeste makes 6% commission on her sales. If her sales for a week total $4,580, what is her commission?

27. Peeler's Jewelry is offering a 30% off sale on all bracelets. How much will you save if you buy a $45.00 bracelet during the sale?

28. How much would an employee pay for a $724 stereo if the employee gets a 15% discount?

29. Misha buys a CD for $14.95. If sales tax is 7%, how much does she pay total?

30. The Pep band made $640 during a fund-raiser. The band spent $400 of the money on new uniforms. What percent of the total did the band members spend on uniforms?

31. Hank gets 10 hours per week to play his video games. Since he made straight "A's" last semester, his parents increased his playing time to 16 hours per week. What percent increase in time does he get?

Chapter 4
Ratios, Proportions, and Scale Drawings

This chapter covers the following Georgia Performance Standards:

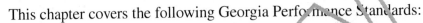

M6G	Geometry	M6G1.c, d, e
M6A	Algebra	M6A1
		M6A2.b, c, d, f, g
M6P	Process Skills	M6P1.c
		M6P2.a, d
		M6P3.d
		M6P4.c, b
		M6P5.a, b, c

4.1   Ratio Problems

In some word problems, you may be asked to express the answer as a **ratio**. Ratios can look like fractions. Numbers must be written in the order they are requested. In the following problem, 8 cups of sugar is mentioned before 6 cups of strawberries. But in the question part of the problem, you are asked for the ratio of STRAWBERRIES to SUGAR. The amount of strawberries IS THE FIRST WORD MENTIONED, so it must be the **top** number of the fraction. The amount of sugar, THE SECOND WORD MENTIONED, must be the **bottom** number of the fraction.

Example 1:   The recipe for jam requires 8 cups of sugar for every 6 cups of strawberries. What is the ratio of strawberries to sugar in this recipe?

$$\frac{\text{First number requested}}{\text{Second number requested}} = \frac{6 \text{ cups strawberries}}{8 \text{ cups sugar}}$$

Answers may be reduced to lowest terms. $\frac{6}{8} = \frac{3}{4}$

Practice writing ratios for the following word problems and reduce to lowest terms. DO NOT CHANGE ANSWERS TO MIXED NUMBERS. Ratios should be left in fraction form.

1. Out of the 248 6th graders, 112 are boys. What is the ratio of boys to the total number of 6th graders?

2. It takes 7 cups of flour to make 2 loaves of bread. What is the ratio of cups of flour to loaves of bread?

3. A skyscraper that stands 620 feet tall casts a shadow that is 125 feet long. What is the ratio of the shadow to the height of the skyscraper?

4. Twenty boxes of paper weigh 520 pounds. What is the ratio of boxes to pounds?

5. The newborn weighs 8 pounds and is 22 inches long. What is the ratio of weight to length?

6. Jack pays $6.00 for 10 pounds of apples. What is the ratio of the price of apples to the pounds of apples?

7. Jordan spends $45 on groceries. Of that total, $23 is for steaks. What is the ratio of steak cost to the total grocery cost?

8. Madison's flower garden measures 8 feet long by 6 feet wide. What is the ratio of length to width?

4.2   Solving Proportions

Two **ratios (fractions)** that are **equal** to each other are called **proportions.** For example, $\frac{1}{4} = \frac{2}{8}$. Read the following example to see how to find a number missing from a proportion.

Example 2: $\dfrac{5}{15} = \dfrac{8}{x}$

Step 1:   To find x, you first multiply the two numbers that are diagonal to each other.

$$\frac{5}{\{15\}} = \frac{\{8\}}{x}$$

$15 \times 8 = 120$ and $5 \times x = 5x$

Therefore, $5x = 120$

Step 2:   Then divide the product (120) by the other number in the proportion (5).

$120 \div 5 = 24$

Therefore, $\dfrac{5}{15} = \dfrac{8}{24}$ **and** $x = 24$.

Practice finding the number missing from the following proportions. First, multiply the two numbers that are diagonal from each other. Then divide by the other number.

1. $\dfrac{2}{5} = \dfrac{6}{x}$

2. $\dfrac{9}{3} = \dfrac{x}{5}$

3. $\dfrac{x}{12} = \dfrac{3}{4}$

4. $\dfrac{7}{x} = \dfrac{3}{9}$

5. $\dfrac{12}{x} = \dfrac{2}{5}$

6. $\dfrac{12}{x} = \dfrac{4}{3}$

7. $\dfrac{27}{3} = \dfrac{x}{2}$

8. $\dfrac{1}{x} = \dfrac{3}{12}$

9. $\dfrac{15}{2} = \dfrac{x}{4}$

10. $\dfrac{7}{14} = \dfrac{x}{6}$

11. $\dfrac{5}{6} = \dfrac{10}{x}$

12. $\dfrac{4}{x} = \dfrac{3}{6}$

13. $\dfrac{x}{5} = \dfrac{9}{15}$

14. $\dfrac{9}{18} = \dfrac{x}{2}$

15. $\dfrac{5}{7} = \dfrac{35}{x}$

16. $\dfrac{x}{2} = \dfrac{8}{4}$

17. $\dfrac{15}{20} = \dfrac{x}{8}$

18. $\dfrac{x}{40} = \dfrac{5}{100}$

4.3   Ratio and Proportion Word Problems

You can use ratios and proportions to solve problems.

Example 3:   A stick one meter long is held perpendicular to the ground and casts a shadow 0.4 meters long. At the same time, an electrical tower casts a shadow 112 meters long. Use ratio and proportion to find the height of the tower.

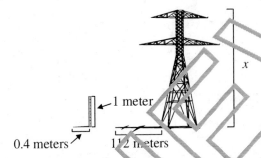

Step 1:   Set up a proportion using the numbers in the problem. Put the shadow lengths on one side of the equation and put the heights on the other side. The 1 meter height is paired with the 0.4 meter length, so let them both be top numbers. Let the unknown height be x.

shadow length   object height

$$\frac{0.4}{112} = \frac{1}{x}$$

Step 2:   Solve the proportion as you did on the previous page.

$$112 \times 1 = 112$$

$$112 \div 0.4 = 280$$

Answer: The tower height is 280 meters.

Use ratio and proportion to solve the following problems.

1. Rudolph can mow a lawn that measures 1000 square feet in 2 hours. At that rate, how long would it take him to mow a lawn 3500 square feet?

2. Faye wants to know how tall her school building is. On a sunny day, she measures the shadow of the building to be 6 feet. At the same time she measures the shadow cast by a 5 foot statue to be 2 feet. How tall is her school building?

3. Out of every 5 students surveyed, 2 listen to country music. At that rate, how many students in a school of 800 listen to country music?

4. Butterfly, a Labrador Retriever, has a litter of 8 puppies. Four are black. At that rate, how many puppies in a litter of 10 would be black?

5. According to the instructions on a bag of fertilizer, 5 pounds of fertilizer are needed for every 100 square feet of lawn. How many square feet will a 25-pound bag cover?

6. A race car can travel 2 laps in 5 minutes. At this rate, how long will it take the race car to complete 100 laps ?

7. If it takes 7 cups of flour to make 4 loaves of bread, how many loaves of bread can you make from 35 cups of flour?

4.4   **Proportional Reasoning**

Proportional reasoning can be used when a selected number of individuals are tagged in a population in order to estimate the total population.

Example 4:   A team of scientists capture, tag, and release 50 deer in a particular national forest. One week later, they capture another 50 deer, and 2 of the deer are ones that were tagged previously. What is the approximate deer population in the national forest?

Solution: Use proportional reasoning to determine the total deer population. You know that 50 deer out of the total deer population in the forest were tagged. You also know that 2 out of those 50 were recaptured. These two ratios should be equal because they both represent a fraction of the total deer population.

$$\frac{50 \text{ deer tagged}}{x \text{ deer tagged}} = \frac{2 \text{ deer tagged}}{50 \text{ deer tagged}}$$

$$2x = 2,500$$
$$x = 1,250 \text{ total deer}$$

Use proportional reasoning to solve the following problems.

1. Dr. Wolf, the biologist, captures 20 fish out of a small lake behind his college. He fastens a marker onto each of these and throws them back into the lake. A week later, he again captures 20 fish. Of these, 2 have markers. How many fish could Dr. Wolf estimate are in the pond?

2. Tawanda draws 20 cards from a box. She marks each one, returns them to the box, and shakes the box vigorously. She then draws 20 more cards and finds that 5 of them are marked. Estimate how many cards are in the box.

3. Maureen pulls 100 pennies out of her money jar, which contains only pennies. She marks each of these, puts them back in the bank, shakes vigorously, and again pulls out 100 pennies. She discovers that 2 of them are marked. Estimate how many pennies are in her money jar.

4. Mr. Kizer has a ten-acre wooded lot. He catches 20 squirrels, tags them, and releases them. Several days later, he catches another 20 squirrels. One of the 20 squirrels has a tag. Estimate the number of squirrels living on Mr. Kizer's ten acres. Assume the squirrels just stay on his property.

5. In Mrs. Stern's 6th grade class, forty percent of the students are wearing blue jeans. If 10 students are wearing blue jeans, how many total students are in Mrs. Stern's class?

4.5   Maps and Scale Drawings

Example 5:   On a map drawn to scale, 5 cm represents 30 kilometers. A line segment connecting two cities is 7 cm long. What distance does this line segment represent?

Step 1:   Set up a proportion using the numbers in the problem. Keep centimeters on one side of the equation and kilometers on the other. The 5 cm is paired with the 30 kilometers, so let them both be top numbers. Let the unknown distance be x.

$$\begin{array}{cc} \text{cm} & \text{km} \\ \dfrac{5}{7} & = & \dfrac{30}{x} \end{array}$$

Step 2:   Solve the proportion as you have previously.
$7 \times 30 = 210$
$210 \div 5 = 42$
Answer: 7 cm represents 42 km.

Sometimes the answer to a scale drawing problem will be a fraction or a mixed number.

Example 6:   On a scale drawing, 2 inches represents 30 feet. How many inches long is a line segment that represents 5 feet?

Step 1:   Set up the proportion as you did above.

$$\begin{array}{cc} \text{inches} & \text{feet} \\ \dfrac{2}{x} & = & \dfrac{30}{5} \end{array}$$

Step 2:   First multiply the two numbers that are diagonal from each other. Then divide by the other number.
$2 \times 5 = 10$
$10 \div 30$ is less than 1 so express the answer as a fraction and reduce.
$10 \div 30 = \frac{10}{30} = \frac{1}{3}$ inch

Set up proportions for each of the following problems and solve.

1. If 2 inches represents 50 miles on a scale drawing, how long would a line segment be that represents 25 miles?

2. On a scale drawing, 2 cm represents 15 km. A line segment on the drawing is 3 cm long. What distance does this line segment represent?

3. On a map drawn to scale, 5 cm represents 250 km. How many kilometers are represented by a line 6 cm long?

4. If 2 inches represents 80 miles on a scale drawing, how long would a line segment be that represents 280 miles?

5. On a map drawn to scale, 5 cm represents 200 km. How long would a line segment be that represents 260 km?

6. On a scale drawing of a house plan, one inch represents 5 feet. How many feet wide is the bathroom if the width on the drawing is 3 inches?

4.6   Using a Scale On a Blueprint

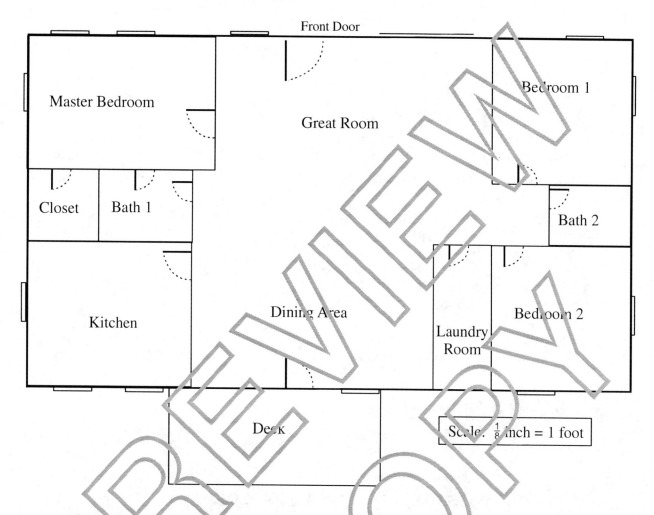

Use a ruler to find the measurements of the rooms on the blueprint above. Convert to feet using the scale. The first problem is done for you.

	long wall		short wall	
	ruler measurement	room measurement	ruler measurement	room measurement
1. Kitchen	$1\frac{3}{4}$ in	14 ft	$1\frac{1}{2}$ in	12 ft
2. Deck				
3. Closet				
4. Bedroom 1				
5. Bedroom 2				
6. Master Bedroom				
7. Bath 1				
8. Bath 2				

Chapter 4 Review

Solve the following proportions.

1. $\dfrac{8}{x} = \dfrac{1}{2}$

2. $\dfrac{2}{5} = \dfrac{x}{10}$

3. $\dfrac{x}{6} = \dfrac{3}{9}$

4. $\dfrac{4}{9} = \dfrac{8}{x}$

5. Out of 100 coins, 45 are in mint condition. What is the ratio of mint condition coins to the total number of coins?

6. The ratio of boys to girls in the seventh grade is 6 : 5. If there are 135 girls in the class, how many boys are there?

7. Twenty out of the total 235 students have straight "A's". What is the ratio of students with straight "A's" to the total number of students?

8. Aunt Bess uses 3 cups of oatmeal to bake 6 dozen oatmeal cookies. How many cups of oatmeal would she need to bake 15 dozen cookies?

9. On a map, 2 centimeters represents 150 kilometers. If a line between two cities measures 5 centimeters, how many kilometers apart are they?

10. When Rick measures the shadow of a yard stick, it is 5 inches. At the same time, the shadow of the tree he would like to chop down is 45 inches. How tall is the tree in yards?

11. If 4 inches represents 8 feet on a scale drawing, how many feet does 6 inches represent?

12. On a map scale, 2 centimeters represents 5 kilometers. If two towns on the map are 20 kilometers apart, how long would the line segment be between the two towns on the map?

13. Jamal wonders how many ants are in his ant farm. He puts a stick in the container, and when he pulls it out, there are 15 ants on it. He gently sprays these ants with a mixture of water and green food coloring, then puts them back into the container. The next day his stick draws 20 ants, 1 of which is green. Estimate how many ants Jamal has.

Chapter 5
Patterns and Problem Solving

This chapter covers the following Georgia Performance Standards:

M6A	Algebra	M6A2.a
M6P	Process Skills	M6P1.a, b, c, d
		M6P2.a, b, c, d
		M6P3.a, c, d
		M6P4.c
		M6P5.a, b, c

5.1   Number Patterns

In each of the examples below, there is a sequence of numbers that follows a pattern. Think of the sequence of numbers like the output for a function. You must find the pattern (or function) that holds true for each number in the sequence. Once you determine the pattern, you can find the next number in the sequence or any number in the sequence.

	Sequence	Pattern	Next Number	20th number in the sequence
Example 1:	$3, 4, 5, 6, 7$	$n + 2$	8	22

In number patterns, the sequence in the output. The input can be the set of whole numbers starting with 1. But you must determine the "rule" or pattern. Look at the table below.

input	sequence
1 →	3
2 →	4
3 →	5
4 →	6
5 →	7

What pattern or "rule" can you come up with that gives you the first number in the sequence, 3, when you input 1? $n + 2$ will work because when $n = 1$, the first number in the sequence $= 3$. Does this pattern hold true for the rest of the numbers in the sequence? Yes, it does. When $n = 2$, the second number in the sequence $= 4$. When $n = 3$, the third number in the sequence $= 5$, and so on. Therefore, $n + 2$ is the pattern. Even without knowing the algebraic form of the pattern, you could figure out that 8 is the next pattern in the sequence. To find the 20th number in the pattern, use $n = 20$ to get 22.

	Sequence	Pattern	Next Number	20th number in the sequence
Example 2:	$1, 4, 9, 16, 25$	n^2	36	400
Example 3:	$-2, -4, -6, -8, -10$	$-2n$	-12	-40

Find the pattern and the next number in each of the sequences below.

	Sequence	Pattern	Next Number	20th number in the sequence
1.	$-2, -1, 0, 1, 2$			
2.	$5, 6, 7, 8, 9$			
3.	$3, 7, 9, 11, 15, 19$			
4.	$-3, -6, -9, -12, -15$			
5.	$3, 5, 7, 9, 11$			
6.	$2, 4, 8, 16, 32$			$1,048,576$
7.	$1, 8, 27, 64, 125$			
8.	$0, -1, -2, -3, -4$			
9.	$2, 5, 10, 17, 26$			
10.	$4, 6, 8, 10, 12$			

5.2   Inductive Reasoning and Patterns

Humans have always observed what happened in the past and used these observations to predict what would happen in the future. This is called **inductive reasoning**. Although mathematics is referred to as the "deductive science," it benefits from inductive reasoning. We observe patterns in the mathematical behavior of a phenomenon, then find a rule or formula for describing and predicting its future mathematical behavior. There are lots of different kinds of predictions that may be of interest.

Example 4:   Nancy is watching her nephew, Drew, arrange his marbles in rows on the kitchen floor. The figure below shows the progression of his arrangement.

Row 1
Row 2
Row 3
Row 4

Assuming this pattern continues, how many marbles would Drew place in a fifth row?

Solution: It appears that Drew doubles the number of marbles in each successive row. In the 4th row he had 8 marbles, so in the 5th row we can predict 16 marbles.

Example 5:   Manuel drops a golf ball from the roof of his high school while Carla video tapes the motion of the ball. Later, the video is analyzed and the results are recorded concerning the height of each bounce of the ball.

What height do you predict for the fifth bounce?

Initial height	1st bounce	2nd bounce	3rd bounce	4th bounce
30 ft	18 ft	10.8 ft	6.48 ft	3.888 ft

To answer this question, we need to be able to relate the height of each bounce to the bounce immediately preceding it. Perhaps the best way to do this is with **ratios** as follows:

$$\frac{\text{Height of 1st bounce}}{\text{Initial bounce}} = 0.6 \qquad \frac{\text{Height of 2nd bounce}}{\text{Height of 1st bounce}} = 0.6$$

$$\frac{\text{Height of 4th bounce}}{\text{Height of 3rd bounce}} = 0.6$$

Since the ratio of the height of each bounce to the bounce before it appears constant, we have some basis for making predictions. Using this, we can reason that the fifth bounce will be equal to 0.6 of the fourth bounce.
Thus we predict the fifth bounce to have a height of $0.6 \times 3.888 = 2.3328$ ft.

Which bounce will be the last one with a height of one foot or greater?

Look at predicted bounce heights until a bounce of less than 1 foot is reached.
The sixth bounce is predicted to be $0.6 \times 2.3328 = 1.399\,768$ ft.
The seventh bounce is predicted to be $0.6 \times 1.399\,768 = 0.839\,808$ ft.
The last bounce with a height greater than 1 ft is predicted to be the sixth one.

Read the following questions carefully. Use inductive reasoning to answer each question. You may wish to make a table or a diagram to help you visualize the pattern in some of the problems.

1. Bob and Alice's older brothers showed then how to design and create a website for their middle school. The first week they had 5 visitors to the site; the second week, they had 10 visitors; and during the third week, they had 20 visitors.
   (A) If current trends continue, how many visitors can they expect in the fifth week?
   (B) How many in the nth week?
   (C) How many weeks will it be before they get more than 500 visitors in a single week?

2. In 1979 (the first year of classes), there were 500 students at Brookstone Middle. In 1989, there were 1000 students. In 1999, there were 2000 students. How many students would you predict at Brookstone in 2009 if this pattern continues (and no new schools are built)?

3. The average combined (math and verbal) SAT score for students at Brookstone High was 1000 in 2001, 1100 in 2002, 1210 in 2003, and 1331 in 2004. Predict the combined SAT score for Brookstone seniors in 2005.

4. Marie has a daylily in her mother's garden. Every Saturday morning in the spring, she measures and records its height in the table below. What height do you predict for Marie's daylily on April 29? (Hint: Look at the *change* in height each week.)

April 1	April 8	April 15	April 22
12 in	18 in	21 in	22.5 in

5. Bob puts a glass of water in the freezer and records the temperature every 15 minutes. The results are displayed in the table below. If this pattern of cooling continues, what will be the temperature at 2:15 P.M.? (Hint: Look at the changes in temperature.)

1:00 P.M.	1:15 P.M.	1:30 P.M.	1:45 P.M.
92° F	60° F	44° F	36° F

Example 6:   Mr. Applegate wants to put desks together in his math class so that students can work in groups. The diagram below shows how he wishes to do it.

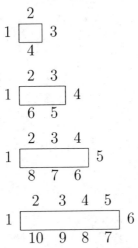

With one table he can seat 4 students, with two tables he can seat 6, with three tables 8, and with four tables 10.

How many students can he seat with 5 tables?

With 5 tables he could seat 5 students along the sides of the tables and 1 student on each end; thus, a total of 12 students could be seated.

Write a rule that Mr. Applegate could use to tell how many students could be seated at n tables. Explain how you got the rule.

For n tables, there would be n students along each of 2 sides and 2 students on the ends (1 on each end); thus, a total of $2n + 2$ students could be seated at n tables.

Example 7:   When he isn't playing football for the Brookstone Bears, Tim designs web pages. A car dealership paid Tim $500 to start a site with photos of its cars. The dealer also agreed to pay Tim $50 for each customer who buys a car first viewed on the website.

Write and explain a rule that tells how much the dealership will pay Tim for the design of the website and the sale of n cars from the website.

Tim's payment will be the initial $500 plus $50 for each sale. Translated into mathematical language, if Tim sells n cars he will paid a total of $500 + 50n$ dollars.

How many cars have to be sold from his site in order for Tim to get $1000 from the dealership?

He earned $500 just by establishing the site, so he only needs to earn an additional $500, which at $50 per car requires the sale of only 10 cars. (Note: Another way to solve this problem is to use the rule found in the first question. In that case, you simply solve the equation $500 + 50n = 1000$ for the variable n).

Example 8: Eric is baking muffins to raise money for the Homecoming dance. He makes 18 muffins with each batch of batter, but he must give one muffin each to his brother, his sister, his dog, and himself each time a batch is finished baking.

Write a rule for the number of muffins Eric produces for the fund-raiser with *n* batches. He bakes 18 muffins with each batch, but only 14 are available for the fund-raiser. Thus with n batches he will produce $14n$ muffins for Homecoming. **The rule = $14n$.**

Use your rule to determine how many muffins he will contribute if he makes 7 batches. The number of batches, n, equals 7. Therefore, he will produce $14 \times 7 = 98$ muffins with 7 batches.

Determine how many batches he must make in order to contribute at least 150 muffins. Ten batches will produce $10 \times 14 = 140$ muffins. Eleven batches will produce $11 \times 14 = 154$ muffins. To produce at least 150 muffins, he must bake at least 11 batches.

Determine how many muffins he would actually bake in order to contribute 150 muffins. Since Eric actually bakes 18 muffins per batch, 11 batches would result in Eric baking $11 \times 18 = 198$ muffins.

Carefully read and solve the problems below. Show your work.

1. Tito is building a picket fence along both sides of the driveway leading up to his house. He will have to place posts at both ends and at every 10 feet along the way because the fencing comes in prefabricated ten-foot sections.

   (A) How many posts will he need for a 180-foot driveway?
   (B) Write and explain a rule for determining the number of posts needed for n ten-foot sections.
   (C) How long of a driveway can he fence with 32 posts?

2. Dakota's beginning pay at his new job is $300 per week. For every three months he continues to work there, he will get a $10 per week raise.

   (A) Write a formula for Dakota's weekly pay after n three-month periods.
   (B) After n years?
   (C) How long will he have to work before his pay gets to $400 per week?

3. Amanda is selling shoes this summer. In addition to her hourly wages, Amanda got a $100 bonus just for accepting the position, and she gets a $2 bonus for each pair of shoes she sells.

   (A) Write and explain a rule that tells how much she will make in bonuses if she sells n pairs of shoes.
   (B) How many pairs of shoes must she sell in order to make $200 in bonuses?

4. The table below displays data relating temperature in degrees Fahrenheit to the number of chirps per minute for a cricket.

Temp (°F)	50	52	55	58	60	64	68
Chirps/min	40	48	60	72	80	96	112

   Write a formula or rule that predicts the number of chirps per minute when the temperature is n degrees.

5.3   Mathematical Reasoning/Logic

The Georgia mathematics curriculum calls for skill development in mathematical **reasoning** or **logic**. The ability to use logic is an important skill for solving math problems, but it can also be helpful in real-life situations. For example, if you need to get to Park Street and the Park Street bus always comes to the bus stop at 3 PM, then you know that you need to get to the bus stop by at least 3 PM. This is a real-life example of using logic, which many people would call "common sense."

There are many different types of statements which are commonly used to describe mathematical principles. However, using the rules of logic, the truth of any mathematical statement must be evaluated. Below is a list of tools used in logic to evaluate mathematical statements.

Logic is the discipline that studies valid reasoning. There are many forms of valid arguments, but we will just review a few here.

A **proposition** is usually a declarative sentence which may be true or false.

An **argument** is a set of two or more related propositions, called **premises**, that provide support for another proposition, called the **conclusion**.

Deductive reasoning is an argument which begins with general premises and proceeds to a more specific conclusion. Most elementary mathematical problems use deductive reasoning.

Inductive reasoning is an argument in which the truth of its premises make it likely or probable that its conclusion is true.

ARGUMENTS

Most of logic deals with the evaluation of the validity of arguments. An argument is a group of statements that includes a conclusion and at least one premise. A premise is a statement that you know is true or at least you assume to be true. Then, you draw a conclusion based on what you know or believe is true in the premise(s). Consider the following example:

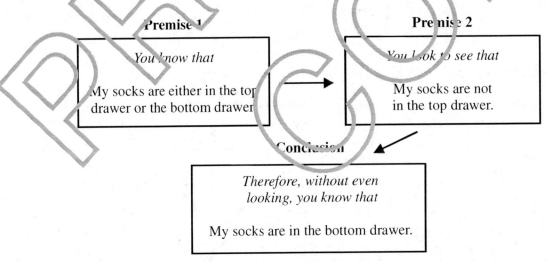

This argument is an example of deductive reasoning, where the conclusion is "deduced" from the premises and nothing else. In other words, if Premise 1 and Premise 2 are true, you don't even need to look in the bottom drawer to know that the conclusion is true.

5.4   Deductive and Inductive Arguments

In general, there are two types of logical arguments: **deductive** and **inductive**. Deductive arguments tend to move from general statements or theories to more specific conclusions. Inductive arguments tend to move from specific observations to general theories.

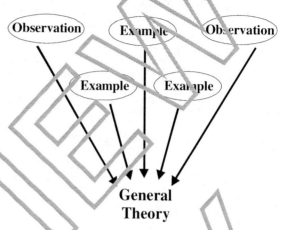

Compare the two examples below:

Deductive Argument

Premise 1	All men are mortal.
Premise 2	Socrates is a man.
Conclusion	Socrates is mortal.

Inductive Argument

Premise 1	The sun rose this morning.
Premise 2	The sun rose yesterday morning.
Premise 3	The sun rose two days ago.
Premise 4	The sun rose three days ago.
Conclusion	The sun will rise tomorrow.

An inductive argument cannot be proved beyond a shadow of a doubt. For example, it's a pretty good bet that the sun will come up tomorrow, but the sun not coming up presents no logical contradiction.

On the other hand, a deductive argument can have logical certainty, but it must be properly constructed. Consider the examples below.

True Conclusion from an Invalid Argument

All men are mortal.
Socrates is mortal.
Therefore Socrates is a man.

Even though the above conclusion is true, the argument is based on invalid logic. Both men and women are mortal. Therefore, Socrates could be a woman.

False Conclusion from a Valid Argument

All astronauts are men.
Julia Roberts is an astronaut.
Therefore, Julian Roberts is a man.

In this case, the conclusion is false because the premises are false. However, the logic of the argument is valid because *if* the premises were true, then the conclusion would be true.

A **counterexample** is an example given in which the statement is true but the conclusion is false when we have assumed it to be true. If we said "All cocker spaniels have blonde hair," then a counterexample would be a red-haired cocker spaniel. If we made the statement, "If a number is greater than 10, it is less than 20," we can easily think of a counterexample, like 35.

Example 9:   Which argument is valid?

If you speed on Hill Street, you will get a ticket.
If you get get a ticket, you will pay a fine.

(A) I paid a fine, so I was speeding on Hill Street.
(B) I got a ticket, so I was speeding on Hill Street.
(C) I exceeded the speed limit on Hill Street, so I paid a fine.
(D) I did not speed in Hill Street, so I did not pay a fine.

Solution: C is valid.
A is incorrect. I could have paid a fine for another violation.
B is incorrect. I could have gotten a ticket for some other violation.
D is incorrect. I could have paid a fine for speeding somewhere else.

Example 10:   Assume the given proposition is true. Then, determine if each statement is true or false.

Given: If a dog is thirsty, he will drink.

(A)   If a dog drinks, then he is thirsty.   T or F
(B)   If a dog is not thirsty, he will not drink.   T or F
(C)   If a dog will not drink, he is not thirsty.   T or F

Solution: A is false. He is not necessarily thirsty; he could just drink because other dogs are drinking or drink to show others his control of the water. This statement is the **converse** of the original. The converse of the statement "If A, then B" is "If B, then A."

B is false. The reasoning from A applies. This statement is the **inverse** of the original. The inverse of the statement "If A, then B" is "If not A, then not B."

C is true. It is the **contrapositive**, or the complete opposite of the original. The contrapositive says "If not B, then not A."

For numbers 1–5, what conclusion can be drawn from each proposition?

1. All squirrels are rodents. All rodents are mammals. Therefore,

2. All fractions are rational numbers. All rational numbers are real numbers. Therefore,

3. All squares are rectangles. All rectangles are parallelograms. All parallelograms are quadrilaterals. Therefore,

4. All Chevrolets are made by General Motors. All Luminas are Chevrolets. Therefore,

5. If a number is even and divisible by three, then it is divisible by six. Eighteen is divisible by six. Therefore,

For numbers 6–9, assume the given proposition is true. Then, determine if the statements following it are true or false.

All squares are rectangles.

6. All rectangles are squares.   T or F
7. All non-squares are non-rectangles. T or F
8. No squares are non-rectangles.   T or F
9. All non-rectangles are non-squares. T or F

Chapter 5 Review

Find the pattern for the following number sequences, and then find the nth number requested.

1. 0, 1, 2, 3, 4   pattern_____   4. 1, 3, 5, 7, 9   25th number_____

2. 0, 1, 2, 3, 4   20th number_____   5. 3, 6, 9, 12, 15   pattern_____

3. 1, 3, 5, 7, 9   pattern_____   6. 3, 6, 9, 12, 15   30th number_____

Justin receives a bill from his internet service provider. The first four months of service are charged according to the table below:

	January	February	March	April
Hours	0	10	5	25
Charge	$4.95	$14.45	$9.70	$28.70

7. Write a formula for the cost of n hours of internet service.

8. What is the greatest number of hours he can get on the internet and still keep his bill under $20.00?

Lisa is baking cookies for the Fall Festival. She bakes 27 cookies with each batch of batter. However, she has a defective oven, which results in 5 cookies in each batch being burnt.

9. Write a formula for the number of cookies available for the festival as a result of Lisa baking n batches of cookies.

10. How many batches does she need in order to produce 300 cookies for the festival?

11. How many cookies (counting burnt ones) will she actually bake?

Chapter 6
Solving One-Step Equations

This chapter covers the following Georgia Performance Standards:

M6A	Algebra	M6A2.f
		M6A3
M6P	Process Skills	M6P1.c
		M6P3.d
		M6P4.a, b

6.1   One-Step Algebra Problems with Addition and Subtraction

You have been solving algebra problems since second grade by filling in blanks. For example, $5 + \underline{\ \ } = 8$. The answer is 3. You can solve the same kind of problems using algebra. The problems only look a little different because the blank has been replaced with a letter. The letter is called a **variable**.

Example 1:   **Arithmetic** $5 + \underline{\ \ } = 14$
                   Algebra $5 + x = 14$

The goal in any algebra problem is to move all the numbers to one side of the equal sign and have the letter (called a **variable**) on the other side. In this problem the 5 and the "x" are on the same side. The 5 is added to x. To move it, do the **opposite** of **add**. The **opposite** of **add** is **subtract**, so subtract 5 from both sides of the equation. Now the problem looks like this:

$$\begin{array}{r} 5 + x = 14 \\ -5 \qquad -5 \\ \hline x = 9 \end{array}$$

To check your answer, put 9 in place of x in the original problem. Does $5 + 9 = 14$? Yes, it does.

Example 2:

$$\begin{array}{r} y - 16 = 27 \\ +16 \quad +16 \\ \hline y = 43 \end{array}$$

Again, the 16 has to move. To move it to the other side of the equation, we do the **opposite** of **subtract**. We **add** 16 to both sides. Check by putting 43 in place of the y in the original problem. Does $43 - 16 = 27$? Yes.

Solve the problems below.

1. $n + 9 = 27$
2. $12 + y = 55$
3. $51 + v = 67$
4. $f + 16 = 31$
5. $5 + x = 23$

6. $15 + x = 24$
7. $w - 14 = 89$
8. $t - 26 = 20$
9. $m - 12 = 17$
10. $c - 7 = 21$

11. $k - 5 = 29$
12. $a + 17 = 45$
13. $d + 26 = 56$
14. $15 + x = 56$
15. $y + 19 = 32$

16. $t - 16 = 28$
17. $m + 14 = 37$
18. $y - 21 = 29$
19. $f + 7 = 31$
20. $h - 12 = 18$

21. $r - 12 = 37$
22. $h - 17 = 22$
23. $x - 37 = 46$
24. $r - 11 = 28$
25. $t - 5 = 52$

6.2   One-Step Algebra Problems with Multiplication and Division

Solving one-step algebra problems with multiplication and division are just as easy as adding and subtracting. Again, you perform the **opposite** operation. If the problem is a **multiplication** problem, you **divide** to find the answer. If it is a **division** problem, you **multiply** to find the answer. Carefully read the examples below, and you will see how easy they are.

Example 3: $4x = 20$ (4x means 4 times x. 4 is the coefficient of x.)

The goal is to get the numbers on one side of the equal sign and the variable x on the other side. In this problem, the 4 and the x are on the same side of the equal sign. The 4 has to be moved over. $4x$ means 4 times x. The opposite of **multiply** is **divide**. If we divide both sides of the equation by 4, we will find the answer.

$4x = 20$ **We need to divide both sides by 4.**

This means divide by 4. $\dfrac{4x}{4} = \dfrac{20}{4}$ We see that $1x = 5$, so $x = 5$.

When you put 5 in place of x in the original problem, it is correct. $4 \times 5 = 20$

Example 4: $\dfrac{y}{4} = 2$

This problem means y divided by 4 is equal to 2. In this case, the opposite of divide is multiply. We need to multiply both sides of the equation by 4.

$4 \times \dfrac{y}{4} = 2 \times 4$ so $y = 8$

When you put 8 in place of y in the original problem, it is correct. $\dfrac{8}{4} = 2$

Solve the problems below.

1. $2x = 14$

2. $\dfrac{w}{5} = 11$

3. $3h = 45$

4. $\dfrac{x}{4} = 36$

5. $\dfrac{x}{3} = 9$

6. $6d = 66$

7. $\dfrac{w}{9} = 3$

8. $7r = 98$

9. $\dfrac{y}{3} = 2$

10. $10y = 30$

11. $\dfrac{r}{4} = 7$

12. $8t = 96$

13. $\dfrac{z}{2} = 15$

14. $\dfrac{n}{9} = 5$

15. $4z = 24$

16. $6d = 84$

17. $\dfrac{t}{3} = 3$

18. $\dfrac{m}{6} = 9$

19. $9p = 72$

20. $5a = 60$

Sometimes the answer to the algebra problem is a **fraction**. Read the example below.

Example 5: $4x = 5$

Problems like this are solved just like the problems above and those on the previous page. The only difference is that the answer is a fraction.

In this problem, the 4 is **multiplied** by x. To solve, we need to divide both sides of the equation by 4.

$4x = 5$ Now **divide** by 4. $\dfrac{4x}{4} = \dfrac{5}{4}$ Now cancel. $\dfrac{4x}{4} = \dfrac{5}{4}$ So $x = \dfrac{5}{4}$

When you put $\dfrac{5}{4}$ in place of x in the original problem, it is correct.

$4 \times \dfrac{5}{4} = 5$ Now cancel. $\longrightarrow$ $4 \times \dfrac{5}{4} = 5$ So $5 = 5$

Solve the problems below. Some of the answers will be fractions. Some answers will be integers.

1. $2x = 3$

2. $4y = 5$

3. $5t = 2$

4. $12b = 144$

5. $9a = 72$

6. $8y = 16$

7. $7x = 21$

8. $4z = 64$

9. $7x = 126$

10. $6p = 10$

11. $2n = 9$

12. $6x = 11$

13. $15m = 180$

14. $5h = 21$

15. $3y = 8$

16. $2t = 10$

17. $3b = 2$

18. $6c = 14$

19. $4d = 3$

20. $5z = 75$

21. $9y = 4$

22. $7d = 12$

23. $2w = 13$

24. $9g = 81$

25. $6a = 18$

26. $2p = 16$

27. $15w = 3$

28. $5x = 13$

6.3   Multiplying and Dividing with Negative Numbers

Example 6: $-3x = 15$

In the problem, -3 is **multiplied** by x. To find the solution, we must do the opposite. The opposite of **multiply** is **divide**. We must both sides of the equation by -3.

$\dfrac{-3x}{-3} = \dfrac{15}{-3}$ Then cancel. $\dfrac{-3x}{-3} = \dfrac{\overset{5}{\cancel{15}}}{\underset{1}{\cancel{-3}}}$ $x = -5$

Example 7: $\dfrac{y}{-4} = -20$

In this problem, y is **divided** by -4. To find the answer, do the opposite. **Multiply** both sides by -4.

$-4 \times \dfrac{y}{-4} = (-20) \times (-4)$ so $y = 80$

Example 8: $-6a = 2$

The answer to an algebra problem can also be a negative fraction.

$\dfrac{\cancel{6}a}{\cancel{6}} = \dfrac{2}{-6}$ ← reduce to get $a = \dfrac{1}{-3}$ or $-\dfrac{1}{3}$

Note: A negative fraction can be written several different ways.

$$\frac{1}{-3} = \frac{-1}{3} = -\frac{1}{3} = -\left(\frac{1}{3}\right)$$

All mean the same thing

Solve the problems below. Reduce any fractions to lowest terms.

1. $2z = -6$

2. $\dfrac{y}{-5} = 20$

3. $-6k = 54$

4. $4x = -24$

5. $\dfrac{t}{7} = -4$

6. $\dfrac{r}{-2} = -10$

7. $9x = 72$

8. $\dfrac{x}{-6} = 3$

9. $\dfrac{w}{-11} = 5$

10. $5y = -35$

11. $\dfrac{x}{-4} = -9$

12. $7t = -49$

13. $-14x = -28$

14. $\dfrac{m}{3} = -12$

15. $\dfrac{c}{-6} = -6$

16. $\dfrac{d}{8} = -7$

17. $\dfrac{y}{-9} = -4$

18. $-15w = -60$

19. $-12v = 36$

20. $-8z = 32$

21. $-4x = -3$

22. $-12y = 7$

23. $\dfrac{a}{-2} = 22$

24. $-18b = 6$

25. $13a = -36$

26. $\dfrac{b}{-2} = -14$

27. $-24c = -6$

28. $\dfrac{y}{-9} = -6$

29. $\dfrac{x}{-23} = -1$

30. $7x = -7$

31. $-9y = -1$

32. $\dfrac{d}{5} = -10$

33. $\dfrac{z}{-13} = -2$

34. $-5c = 45$

35. $2d = -3$

36. $-8d = -12$

37. $-24w = 9$

38. $-6p = 42$

39. $-9a = -18$

40. $\dfrac{p}{-2} = 15$

6.4   Variables with a Coefficient of Negative One

The answer to an algebra problem should not have a negative sign in front of the variable. For example, the problem $-x = 5$ is not completely solved. Study the examples below to learn how to finish solving this problem.

Example 9: $-x = 5$

$-x$ means the same thing as $-1x$ or -1 times x. To solve this problem, **multiply** both sides by -1.

$(-1)(-1x) = (-1)(5)$ so $x = -5$

Example 10: $-y = 3$ Solve the same way.

$(-1)(-y) = (-1)(-3)$ so $y = -3$

Solve the following problems.

1. $-w = 14$

2. $-a = 20$

3. $-x = -15$

4. $-x = -25$

5. $-y = -16$

6. $-t = 62$

7. $-p = -34$

8. $-m = -61$

9. $-w = 17$

10. $-v = -9$

11. $-k = 13$

12. $-q = 7$

Chapter 6 Review

Solve the following one-step algebra problems.

1. $5y = -25$

2. $x + 4 = 24$

3. $d - 11 = 14$

4. $\frac{a}{6} = -8$

5. $-t = 2$

6. $-14b = 12$

7. $\frac{c}{-10} = -3$

8. $z - 15 = -19$

9. $-13d = 4$

10. $\frac{x}{-14} = 2$

11. $-4k = -12$

12. $y + 13 = 27$

13. $15 + h = 4$

14. $14p = 2$

15. $\frac{b}{4} = 11$

16. $p - 26 = 12$

17. $x + (-2) = 5$

18. $m + 17 = 27$

19. $\frac{k}{-4} = 13$

20. $-18a = -7$

21. $21t = -7$

22. $z - (-9) = 14$

23. $23 + w = 28$

24. $n - 35 = -16$

25. $-t = 26$

26. $19 + f = -9$

27. $\frac{w}{11} = 3$

28. $-7y = 28$

29. $x + 23 = 20$

30. $z - 12 = -7$

31. $16 + g = 40$

32. $\frac{m}{-3} = -9$

33. $d + (-6) = 17$

34. $-p = 47$

35. $k - 16 = 5$

36. $9y = -3$

37. $-2z = -36$

38. $10h = 12$

39. $w - 16 = 4$

40. $y + 10 = -8$

Chapter 7
Introduction to Writing and Graphing Equations

This chapter covers the following Georgia Performance Standards:

M6A	Algebra	M6A2.e, f
M6P	Process Skills	M6P1.b, c
		M6P3.d
		M6P4.a, b,
		M6P5.a, b, c

7.1   Graphing Simple Linear Equations

The Cartesian plane is used to graph the solution set for an equation. Any equation with two variables that are both to the first power is called a **linear equation.** The graph of a linear equation will always be a straight line.

Example 1:   Graph the solution set for $2x + y = 0$.

Step 1:   Make a list of some pairs of numbers that will work in the equation.

$$2x + y = 0$$

$2(1) - 2 = 2 - 2 = 0$	$(1, -2)$
$2(2) - 4 = 4 - 4 = 0$	$(2, -4)$
$2(3) - 6 = 6 - 6 = 0$	$(3, -6)$
$0 + 0 = 0$	$(0, 0)$

ordered pair solutions

Step 2:   Plot these points on a Cartesian plane.

Step 3:   By passing a line through these points, we graph the solution set for $2x + y = 0$. This means that every point on the line is a solution to the equation $2x + y = 0$. The point $(0, 0)$ is a solution, so the line must pass through the point $(0, 0)$.

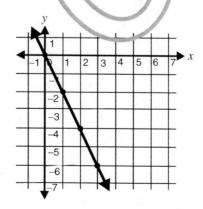

Make a table of solutions for each linear equation below. Then plot the ordered pair solutions on graph paper. Draw a line through the points. (Remember, the points must line up in a straight line.)

1. $-x + y = 0$

2. $y = 4x$

3. $y = -3x$

4. $x = y$

5. $3x = y$

6. $7x - y = 0$

Example 2:   Graph the equation $y = \frac{2}{3}x$.

Step 1:   This equation has 2 variables, both to the first power, so we know the graph will be a straight line. Substitute some numbers for x or y to find pairs of numbers that satisfy the equation. For the above equation, it will be easier to substitute values of x in order to find the corresponding value for y. Record the values for x and y in a table.

If x is -3, y would be -2
If x is 0, y would be 0
If x is 3, y would be 2
If x is 6, y would be 4

x	y
-3	-2
0	0
3	2
6	4

Step 2:   Graph the ordered pairs, and draw a line through the points.

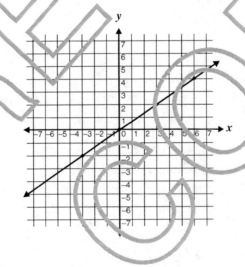

Find pairs of numbers that satisfy the equations below, and graph the line on graph paper.

1. $y = -\frac{2}{3}x$

2. $1.5x = y$

3. $-x = y$

4. $y = 0.5x$

5. $\frac{4}{5}x = y$

6. $y = \frac{3}{4}x$

7. $x = 4y$

8. $2x = 3y$

9. $x + 2y = 0$

7.2   Understanding Slope

The slope of a line refers to how steep a line is. Slope is also defined as the rate of change. When we graph a line using ordered pairs, we can easily determine the slope. Slope is often represented by the letter m.

> The formula for slope of a line is: $m = \dfrac{y_2 - y_1}{x_2 - x_1}$ or $\dfrac{\text{rise}}{\text{run}}$

Example 3:   What is the slope of the following line that passes through the ordered pairs $(-4, -3)$ and $(1, 3)$?

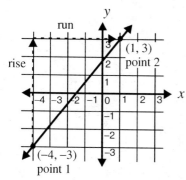

y_2 is 3, the y-coordinate of point 2.
y_1 is -3, the y-coordinate of point 1.
x_2 is 1, the x-coordinate of point 2.
x_1 is -4, the x-coordinate of point 1.

Use the formula for slope given above: $m = \dfrac{3 - (-3)}{1 - (-4)} = \dfrac{6}{5}$

The slope is $\frac{6}{5}$. This shows us that we can go up 6 (rise) and over 5 to the right (run) to find another point on the line.

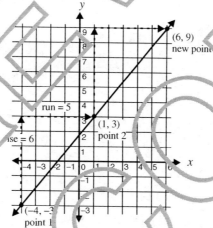

Example 4.   Find the slope of a line through the points $(-2, 3)$ and $(1, -2)$. It doesn't matter which pair we choose for point 1 and point 2. The answer is the same.

Let point 1 be $(-2, 3)$
Let point 2 be $(1, -2)$
slope $= \dfrac{(y_2 - y_1)}{(x_2 - x_1)} = \dfrac{-2 - 3}{1 - (-2)} = \dfrac{-5}{3}$
When the slope is negative, the line will slant left. For this example, the line will go **down** 5 units and then over 3 units to the **right**.

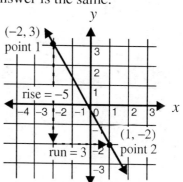

Example 5:   What is the slope of a line that passes through $(1, 1)$ and $(3, 1)$?

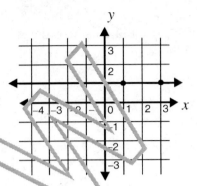

$$\text{slope} = \frac{1 - 1}{3 - 1} = \frac{0}{2} = 0$$

When $y_2 - y_1 = 0$, the slope will equal 0, and the line will be horizontal.

Example 6:   What is the slope of a line that passes through $(2, 1)$ and $(3, 1)$?

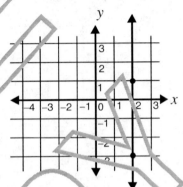

$$\text{slope} = \frac{-3 - 1}{2 - 2} = \frac{4}{0} = \text{undefined}$$

When $x_2 - x_1 = 0$, the slope is undefined and the line will be vertical.

The following lines summarize what we know about slope.

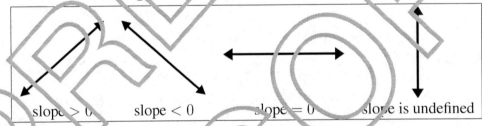

slope > 0 slope < 0 slope $= 0$ slope is undefined

Find the slope of the line that goes through the following pairs of points. Use the formula slope $= \dfrac{y_2 - y_1}{x_2 - x_1}$. Then, using graph paper, graph the line through the two points, and label the rise and run. (See Examples 3 and 4).

1. $(2, 3)$ $(4, 5)$

2. $(1, 3)$ $(2, 5)$

3. $(-1, 2)$ $(4, 1)$

4. $(1, -2)$ $(4, -2)$

5. $(3, 0)$ $(3, 4)$

6. $(3, 2)$ $(-1, 8)$

7. $(4, 3)$ $(2, 4)$

8. $(2, 2)$ $(1, 5)$

9. $(3, 4)$ $(1, 2)$

10. $(3, 2)$ $(3, 6)$

11. $(6, -2)$ $(3, -2)$

12. $(1, 2)$ $(3, 4)$

13. $(-2, 1)$ $(-4, 3)$

14. $(5, 2)$ $(4, -1)$

15. $(1, -3)$ $(-2, 4)$

16. $(2, -1)$ $(3, 5)$

7.3   Graphing Linear Data

We relate many types of data by a constant ratio. As you learned on the previous page, this type of data is linear. The slope of the line described by linear data is the ratio between the data. Plotting linear data with a constant ratio can be helpful in finding additional values.

Example 7:   A department store prices socks per pair. Each pair of socks costs $0.75. Plot pairs of socks versus price on a Cartesian plane.

Step 1:   Since the price of the socks is constant, you know that one pair of socks costs $0.75, 2 pairs of socks cost $1.50, 3 pairs of socks cost $2.25, and so on. Make a list of a few points.

Pair(s) x	Price y
1	0.75
2	1.50
3	2.25

Step 2:   Plot these points on a Cartesian plane, and draw a straight line through the points.

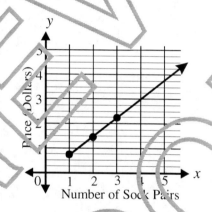

Example 8:   What is the slope of the data in the example above? What does the slope describe?

Solution: You can determine the slope either by the graph or by the data points. For this data, the slope is .75. Remember, slope is rise/run. For every $0.75 going up the y-axis, you go across one pair of socks on the x-axis. The slope describes the price per pair of socks.

Example 9:   Use the graph created in above example to answer the following questions. How much would 5 pairs of socks cost? How many pairs of socks could you purchase for $3.00? Extending the line gives useful information about the price of additional pairs of socks.

Solution 1:   The line that represents 5 pairs of socks intersects the data line at $3.75 on the y-axis. Therefore, 5 pairs of socks would cost $3.75.

Solution 2:   The line representing the value of $3.00 on the y-axis intersects the data line at 4 on the x-axis. Therefore, $3.00 will buy exactly 4 pairs of socks.

Use the information given to make a line graph for each set of data, and answer the questions related to each graph.

1. The diameter of a circle versus the circumference of a circle is a constant ratio. Use the data given below to graph a line to fit the data. Extend the line, and use the graph to answer the next question.

Circle

Diameter	Circumference
4	12.56
5	15.70

2. Using the graph of the data in question 1, estimate the circumference of a circle that has a diameter of 3 inches.

3. If the circumference of a circle is 3 inches, about how long is the diameter?

4. What is the slope of the line you graphed in question 1?

5. What does the slope of the line in question 4 describe?

6. The length of a side on a square and the perimeter of a square are constant ratios to each other. Use the data below to graph this relationship.

Square

Length of side	Perimeter
2	8
3	12

7. Using the graph from question 6, what is the perimeter of a square with a side that measures 4 inches?

8. What is the slope of the line graphed in question 6?

9. Conversions are often constant ratios. For example, converting from pounds to ounces follows a constant ratio. Use the data below to graph a line that can be used to convert pounds to ounces.

Measurement Conversion

Pounds	Ounces
2	32
4	64

10. Use the graph from question 9 to convert 40 ounces to pounds.

11. What does the slope of the line graphs for question 9 represent?

12. Graph the data below, and create a line that shows the conversion from weeks to days.

Time

Weeks	Days
1	7
2	14

13. About how many days are in $2\frac{1}{2}$ weeks?

Chapter 7 Review

Answer the following questions about graphing linear equations.

1. Graph the solution set for the linear equation: $\frac{1}{3}x = y$ on your own graph paper.

2. What is the slope of the line $y = -\frac{1}{2}x$?

3. Paulo turns on the oven to preheat it. After one minute, the oven temperature is $200°$. After 2 minutes, the oven temperature is $325°$.

   Oven Temperature

Minutes	Temperature
1	200°
2	325°

   Assuming the oven temperature rises at a constant rate, write an equation that fits the data.

4. Write an equation that fits the data given below. Assume the data is linear.

   Plumber Charges per Hour

Hour	Charge
1	$170
2	$220

Chapter 8
Data Interpretation

This chapter covers the following Georgia Performance Standards:

M6D	Data Analysis and Probability	M6D1.a, b, c, d, e
M6P	Process Skills	M6P1.b
		M6P3.d
		M6P4.c
		M6P5.a, b, c

8.1   Tally Charts and Frequency Tables

Large lists can be tallied in a chart. To make a **tally chart**, record a tally mark in a chart for each time a number is repeated. To make a **frequency table**, count the times each number occurs in the list, and record the frequency.

Example 1:   The age of each student in grades 6–8 in a local middle school are listed below. Make a tally chart and a frequency table for each age.

Student Ages grades 6–8						
10	11	11	12	14	12	11
13	13	13	12	14	11	12
12	14	12	10	15	11	13
12	10	12	11	12	13	12
13	12	13	12	11	10	13
14	14	11	15	12	13	14
12	11	14	12	11	13	

TALLY CHART	
Age	Tally
10	IIII
11	HHH HHH
12	HHH HHH HHH
13	HHH HHH
14	HH II
15	II

FREQUENCY TABLE	
Age	Frequency
10	4
11	10
12	15
13	10
14	7
15	2

Make a chart to record tallies and frequencies for the following problems.

1. The sheriff's office monitors the speed of cars traveling on Turner Road for one week. The following data is the speed of each car that travels Turner Road during the week. Tally the data in 10 miles per hour (mph) increments starting with 10-19 mph, and record the frequency in a chart.

Car Speed, mph									
45	52	47	35	48	50	51	43	52	41
40	51	32	24	55	41	32	33	45	
36	39	49	52	34	28	39	47	56	
29	15	63	42	35	42	58	59	35	
39	41	25	34	22	16	40	31	55	
55	10	46	38	50	52	48	36	65	
21	32	36	41	52	49	45	32	20	

Speed	Tally	Frequency
10-19		
20-29		
30-39		
40-49		
50-59		
60-69		

2. The following data gives final math averages for Ms. Kirby's class. In her class, an average of 90–100 is an A, 80–89 is B, 70–79 is a C, 60–69 is a D, and an average below 60 is an F. Tally and record the frequency of A's, B's, C's D's, and F's.

Final Math Averages

85	92	87	62	75	84	96	52	31	79
45	77	98	75	71	79	85	82	86	76
87	74	76	68	93	77	65	84	89	
79	65	77	82	86	84	92	60	65	
99	75	88	74	79	80	63	84	69	
87	90	75	81	73	69	73	75	75	

Grade	Tally	Frequency
A		
B		
C		
D		
F		

8.2   Histograms

A **histogram** is a bar graph of the data in a frequency table.

Example 2:   Draw a histogram for the customer sales data presented in the frequency table.

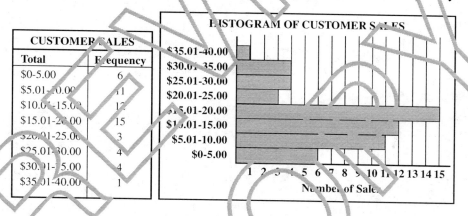

CUSTOMER SALES	
Total	Frequency
$0-5.00	6
$5.01-10.00	11
$10.01-15.00	17
$15.01-20.00	15
$20.01-25.00	3
$25.01-30.00	4
$30.01-35.00	4
$35.01-40.00	1

Use the frequency charts that you filled in the previous section to draw histograms for the same data.

1.
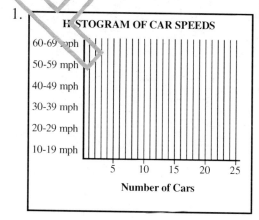

2.
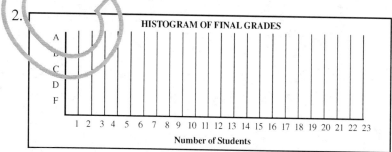

8.3   Reading Tables

A **table** is a concise way to organize large quantities of information using rows and columns. **Read each table carefully, and then answer the questions that follow.**

Some employers use a tax table like the one below to figure how much Federal Income Tax should be withheld from a single person paid weekly. The number of withholding allowances claimed is also commonly referred to as the number of deductions claimed.

Federal Income Tax Withholding Table SINGLE Persons – WEEKLY Payroll Period					
If the wages are –		And the number of withholding allowances claimed is –			
		0	1	2	3
At least	But less than	The amount of income tax to be withheld is –			
$250	260	31	23	16	9
$260	270	32	25	17	10
$270	280	34	26	19	12
$280	290	35	28	20	13
$290	300	37	29	22	15

1. David is single, claims 2 withholding allowances, and earned $275 last week. How much Federal Income Tax was withheld?

2. Cecily claims 0 deductions and she earned $297 last week. How much Federal Income Tax was withheld?

3. Sherri claims 3 deductions and earned $268 last week. How much Federal Income Tax was withheld from her check?

4. Mitch is single and claims 1 allowance. Last week, he earned $291. How much was withheld from his check for Federal Income Tax?

5. Ginger earns $275 this week and claims 0 deductions. How much Federal Income Tax is withheld from her check?

6. Bill is single and earns $263 per week. He claims 1 withholding allowance. How much Federal Income Tax is withheld each week?

8.4   Bar Graphs

Bar graphs can be either vertical or horizontal. There may be just one bar or more than one bar for each interval. Sometimes each bar is divided into two or more parts. In this section, you will work with a variety of bar graphs. Be sure to read all titles, keys, and labels to completely understand all the data that is presented. **Answer the questions about each graph.**

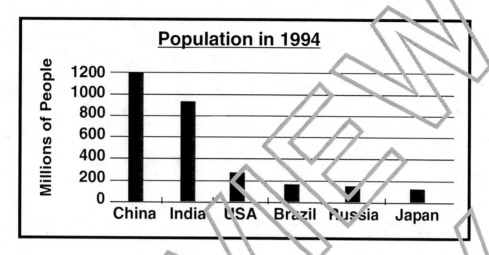

1. Which country has over 1 billion people?

2. How many countries have fewer than 200,000,000 people?

3. How many more people does India have than Japan?

4. If you added together the populations of the USA, Brazil, Russia, and Japan, would it come closer to the population of India or China?

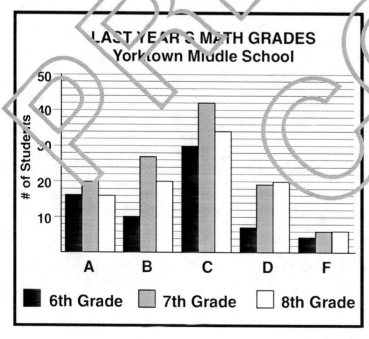

5. How many of last year's 6th graders made C's in math?

6. How many more math students made B's in the 7th grade than in the 8th grade?

7. Which letter grade occurs the most number of times in the data?

8. How many 8th graders took math last year?

9. How many students made A's in math last year?

8.5   Line Graphs

The line graphs below are shown with a globe marking the lines of latitude to make the line graphs more understandable. Study the line graphs below, and then answer the questions that follow.

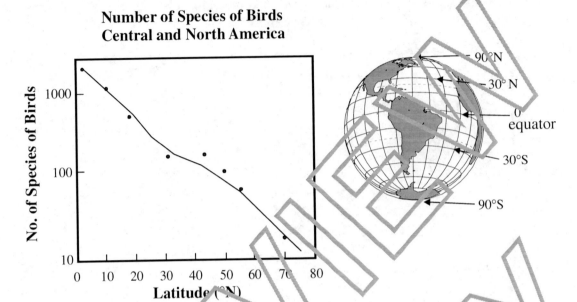

**Number of Species of Birds
Central and North America**

After reading the graph above, label each of the following statements as true or false.

1. There are more species of birds at the North Pole than at the equator.

2. There are more species of birds in Mexico than in Canada.

3. As the latitude increases, the number of species of birds decreases.

4. At 30°N there are over 100 species of birds.

5. The warmer the climate, the fewer kinds of birds there are.

These true or false statements, 6–10, refer to the graph on the left.

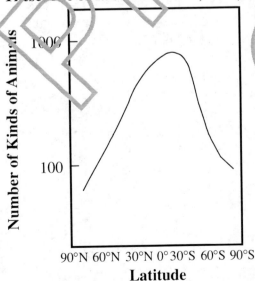

6. The further north and south you go from the equator, the greater the variety of animals there are.

7. The closer you get to the equator, the greater the variety of animals there are.

8. There are fewer kinds of animals at 30°S than at 60°S latitude.

9. The number of kinds of animals increases as the latitude increases.

10. The number of kinds of animals increases at the poles.

8.6   Circle Graphs

Circle graphs represent data expressed in percentages of a total. The parts in a circle graph should always add up to 100%. Circle graphs are sometimes called **pie graphs** or **pie charts**.

To figure the value of a percent in a circle graph, multiply the percent by the total. Use the circle graphs below to answer questions. The first question is worked for you as an example.

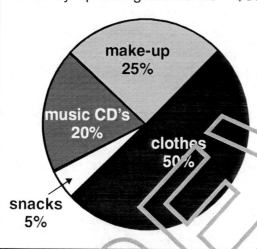

Tina's Monthly Spending Habits

Monthly Spending Allowance = $80

make-up 25%

music CD's 20%

clothes 50%

snacks 5%

1. How much does Tina spend each month on music CD's?

   $80 × 0.20 = $16.00

   ___$16.00___

2. How much does Tina spend each month on make-up?

   _____

3. How much does Tina spend each month on clothes?

   _____

4. How much does Tina spend each month on snacks?

   _____

Fill in the following chart.

Favorite Activity	Number of Students
5. watching TV	
6. talking on the phone	
7. playing video games	
8. surfing the Internet	
9. playing sports	
10. reading	

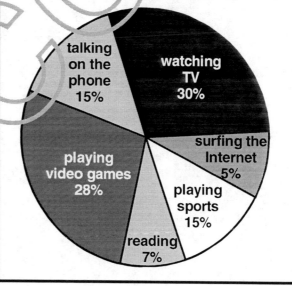

Favorite After-school Activities

Survey of 1000 students (ages 10-15)

talking on the phone 15%

watching TV 30%

surfing the Internet 5%

playing video games 28%

playing sports 15%

reading 7%

8.7   Pictographs

Pictographs represent data using symbols. The **key** or **legend** tells what the symbol stands for. Before answering questions about the graph, be sure to read the title, key, and horizontal and vertical labels.

Number of Military Officers by Branch of Service

U.S. Bureau of the Census, 1994

Key: Each military symbol = 10,000 officers

Answer each question below.

1. How many military officers are in the Marines?

2. Which branch of the service has the fewest number of officers?

3. Are there more Air Force officers or Army officers?

4. How many military officers are there in all?

5. How many officers do the Navy and Marines have altogether?

6. How many more Navy officers are there than Marine officers?

7. How many Army and Coast Guard officers are there altogether?

WORLD PRODUCTION AND CONSUMPTION OF PETROLEUM

● Production   ○ Consumption

Each symbol signifies one million barrels per day

1. What part of the world consumes the most oil?

2. How many parts of the world produce more oil than they consume?

3. How many parts of the world consume more oil than they produce?

4. Which part of the world produces three times more oil than it consumes?

5. How many more barrels of oil does Europe produce daily than Africa?

6. How many barrels does the Middle East produce every day?

7. Which areas of the world consume more than twice as much oil as they produce?

8.8   Graphing Data

Use the given data to make a graph. Then answer the questions that follow.

1. Make a line graph of the following data in the space provided.

Long distance telephone charges

0 minutes	$5.00
10 minutes	$6.00
20 minutes	$7.00
60 minutes	$11.00

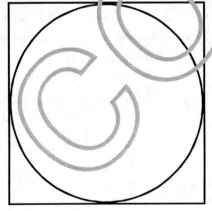

Charges / Minutes

2. What are the charges for 30 minutes of long distance?

3. How many minutes of long distance would be billed for $10.00?

4. What would be the charges for 65 minutes of long distance?

5. A dartboard has been divided into wedges according to the following color percentages. Draw the dartboard in the circle provided and indicate the color of each wedge.

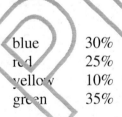

blue	30%
red	25%
yellow	10%
green	35%

6. If you threw the dart at the board without looking, which color is it most likely to hit?

7. Which color would be the hardest to hit?

8.9   Collecting Data Through Surveys

One type of data collection is a **survey**. There are three main types of surveys, mail surveys, telephone surveys, and personal interviews.

Mail surveys are surveys that are sent out to the participants. They usually consist of a **questionnaire**, which is a set of questions. The participants must fill out the surveys and mail it back to the experimenter.

Telephone surveys are surveys that are done over the phone. The person performing the study calls the participant on the phone and asks questions from a questionnaire that was done before the phone call was made.

Personal interviews are surveys that take place in person. The experimenter asks the participants face-to-face questions from a questionnaire. This is the most effective way to use a survey. Participants are less likely to lie to the experimenter, and personal interviews have the best **response rate**, which is the percentage of how many participants actually answered the questions in the survey.

When conducting a survey, you usually cannot get the entire population that you would like to study. The **population** is all of the people that you want to study. For example, if you want to design an experiment where you examine the study habits of 11- and 12-year old students, then your population would be every 11- and 12-year old student in the United States. **Samples** are units selected to study from the population. Taking the previous example, the sample for your experiment might be all of the 11- and 12-year old students in your school.

Chapter 8 Review

Use the data given to answer the questions that follow.

The 6th grade did a survey on the number of pets each student had at home. The following give the data produced by the survey.

NUMBER OF PETS PER STUDENT
0 2 6 2 1 0 4 2 3 3 0 2 5 5 1 4 2 0 5 2 3 3 4 3 6 2
5 1 2 3 5 6 3 2 2 5 2 3 4 3 0 4 4 1 2 4 5 7 6 1 4 7

1. Fill in the frequency table.

Number of Pets	Frequency

2. Fill in the histogram.

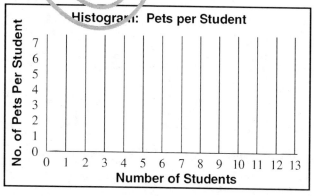

KNIGHTS BASKETBALL Points Scored				
Player	game 1	game 2	game 3	game 4
Joey	5	2	4	8
Jason	10	8	10	12
Brandon	2	6	5	6
Ned	1	3	6	2
Austin	0	4	7	8
David	7	2	9	4
Zac	8	6	7	4

3. How many points do the Knights basketball team score in game 1?

4. How many more points does David score in game 3 than in game 1?

5. How many points does Jason score in the first 4 games?

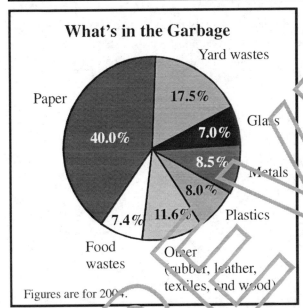

What's in the Garbage

Yard wastes 17.5%
Paper 40.0%
Glass 7.0%
Metals 8.5%
8.0%
Plastics
7.4% 11.6%
Food wastes
Other (rubber, leather, textiles, and wood)

Figures are for 2004.

6. In 2004, the United States produced 160 million metric tons of garbage. According to the pie chart, how much glass was in the garbage?

7. Out of the 160 million metric tons of garbage, how much was glass, plastic, and metal?

8. If in 2006, the garbage reaches 200 million metric tons, and the percentage of wastes remains the same as in 2004, how much food will be in the 2006 garbage?

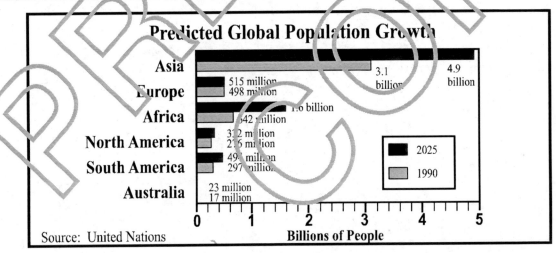

Predicted Global Population Growth

Asia   3.1 billion   4.9 billion
Europe   515 million / 498 million
Africa   1.6 billion / 642 million
North America   312 million / 276 million
South America   492 million / 297 million
Australia   23 million / 17 million

2025
1990

0   1   2   3   4   5
Billions of People

Source: United Nations

9. By how many is Asia's population predicted to increase between 1990 and 2025?

10. In 1990, how much larger was Africa's population than Europe's?

11. Where is the population expected to more than double between 1990 and 2025?

12. In the space below, draw a line graph showing a population increasing over time.

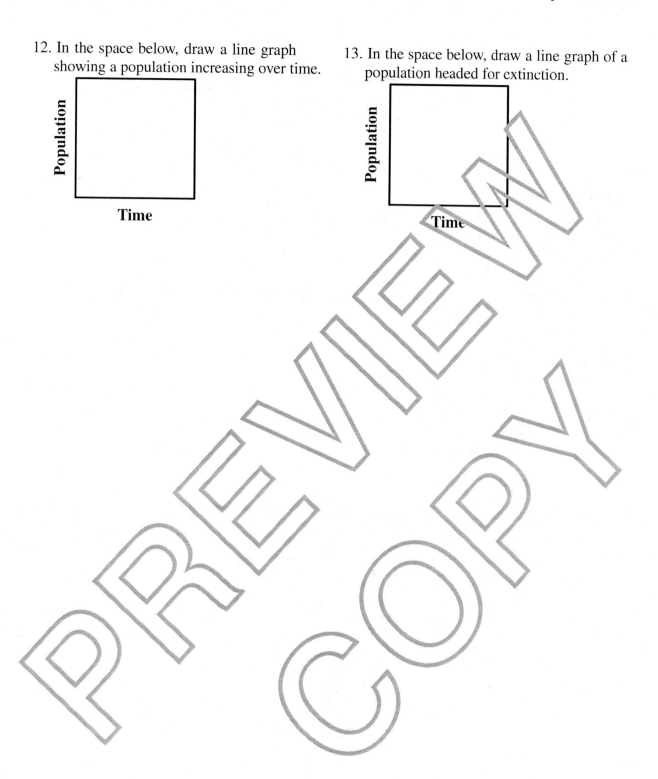

13. In the space below, draw a line graph of a population headed for extinction.

Chapter 9
Probability

This chapter covers the following Georgia Performance Standards:

M6D	Data Analysis and Probability	M6D2.a, b, c
M6P	Process Skills	M6P1.b
		M6P2.b, c
		M6P3.d
		M6P4.c
		M6P5.a, b, c

9.1 Probability

Probability is the chance something will happen. We express probability as a fraction, a decimal, a percent; it can also be written out in words.

Example 1: Billy has 3 red marbles, 5 white marbles, and 4 blue marbles on the floor. His cat comes along and bats one marble under the chair. What is the **probability** it is a red marble?

Step 1: The number of red marbles, 3, will be on top of the fraction.

Step 2: The total number of marbles, 12, will be on the bottom of the fraction. The answer may be expressed in lowest terms. $\frac{3}{12} = \frac{1}{4}$. Expressed as a decimal, $\frac{1}{4} = .25$, as a percent, $\frac{1}{4} = 25\%$, and written out in words, $\frac{1}{4}$ is one out of four.

Example 2: Determine the probability that the pointer will stop on a shaded wedge or the number 1.

Step 1: Count the number of possible wedges that the spinner can stop on to satisfy the above problem. There are 5 wedges that satisfy it (4 shaded wedges and one number 1). The top number of the fraction is 5.

Step 2: Count the total number of wedges, 7. The bottom number of the fraction is 7. The probability that the pointer will stop on a shaded wedge or the number 1 is $\frac{5}{7}$ or five out of seven.

Example 3: Refer to the spinner in example 2. If the pointer stops on the number 7, what is the probability that it will **not** stop on 7 on the next spin?

Step 1: Ignore the information that the pointer stopped on the number 7 on the previous spin. The probability of the next spin does not depend on the outcome of the previous spin. Simply find the probability that the spinner will not stop on 7. Remember, if P is the probability of an event occurring, $1 - P$ is the probability of an event not occurring. In this example, the probability of the spinner landing on 7 is $\frac{1}{7}$.

Step 2: The probability that the spinner will not stop on 7 is $1 - \frac{1}{7}$ which equals $\frac{6}{7}$. The answer is $\frac{6}{7}$ or **six out of seven**.

Find the probability of the following problems. Express the answer as a ratio.

1. A computer chooses a random number between 1 and 50. What is the probability of your guessing the same number that the computer chooses in 1 try?

2. There are 24 candy-coated chocolate pieces in a bag. Eight have defects in the coating that can be seen only with close inspection. What is the probability of pulling out a defective piece without looking?

3. Seven sisters have to choose which day each will wash the dishes. They put equal-sized pieces of paper, each labeled with a day of the week in a hat. What is the probability that the first sister who draws will choose a weekend day?

4. For his garden, Clay has a mixture of 12 white corn seeds, 24 yellow corn seeds, and 16 bicolor corn seeds. If he reaches for a seed without looking, what is the probability that Clay will plant a bicolor corn seed first?

5. Mom just got a new department store credit card in the mail. What is the probability that the last digit is an odd number?

6. Alex has a paper bag of cookies that includes 8 chocolate chip, 4 peanut butter, 6 butterscotch chip, and 12 ginger. Without looking, his friend John reaches in the bag for a cookie. What is the probability that the cookie is peanut butter?

7. An umpire at a little league baseball game has 14 balls in his pockets. Five of the balls are brand A, 6 are brand B, and 3 are brand C. What is the probability that the next ball he throws is a brand C ball?

8. What is the probability that the spinner arrow will land on an even number?

9. Using the spinner above, what is the probability that the spinner will not stop on a shaded wedge or on the number 2 on the third spin?

10. A company is offering 1 grand prize, 3 second place prizes, and 25 third place prizes based on a random drawing of contest entries. If you entered one of the 500 total entries, what is the probability you will win a third place prize?

11. In the contest problem above, what is the probability that you will win the grand prize or a second place prize?

12. A box of a dozen doughnuts has 3 lemon cream-filled, 5 chocolate cream-filled, and 4 vanilla cream-filled. What is the probability of picking a lemon cream-filled?

9.2   More Probability

Example 4:   You have a cube with one number, 1,2,3,4,5 and 6 painted on each face of the cube. What is the probability that if you throw the cube 3 times, you will get the number 2 each time?

If you roll the cube once, you have a 1 in 6 chance of getting the number 2. If you roll the cube a second time, you again have a 1 in 6 chance of getting the number 2. If you roll the cube a third time, you again have a 1 in 6 chance of getting the number 2. The probability of rolling the number 2 three times in a row is:

$$\frac{1}{6} \times \frac{1}{6} \times \frac{1}{6} = \frac{1}{216}$$

Find the probability that each of the following events will occur.

There are 10 balls in a box, each with a different digit on it: 0, 1, 2, 3, 4, 5, 6, 7, 8, & 9. A ball is chosen at random and then put back in the box.

1. What is the probability that if you pick out a ball 3 times, you will get number 7 each time?

2. What is the probability you will pick a ball with 5, then 9, and then 3?

3. What is the probability that if you pick out a ball 4 times, you will always get an odd number?

4. A couple has 4 children ages 9, 6, 4, and 1. What is the probability that they are all girls?

There are 26 letters in the alphabet, allowing a different letter to be on each of 26 cards. The cards are shuffled. After each card is chosen at random, it is put back in the stack of cards, and the cards are shuffled again.

5. What is the probability that when you pick 3 cards, you would draw first a "y", then and "e", and then an "s"?

6. What is the probability that you would draw 4 cards and get the letter "z" each time?

7. What is the probability that you draw twice and get a letter in the word "random" both times?

8. If you flip a coin 3 times, what is the probability you will get heads every time?

9. Marie is clueless about 4 of her multiple-choice answers. The possible answers are A, B, C, D, E, or F. What is the probability that she will guess all four answers correctly?

9.3 Simulations

A **simulation** is usually generated by a computer program. It automatically produces the results of an experiment. To find probabilities, the simulation generates the results from a series of trials. The probabilities that are found from simulations are experimental and are not always accurate.

Example 5: The chart below represents a computer simulation. It shows the frequencies of the results of flipping two coins. The two coins were flipped at the same time 100 times.

Outcome	TT	TH	HT	HH
Frequency	23	35	23	19

Find the theoretical probability of flipping one tail and one head, and find the experimental probability of flipping one tail and one head based on the computer simulation, then compare the two values.

Step 1: Find the theoretical probability. The probability of flipping a tail with coin one is $\frac{1}{2}$, and the probability of flipping a head with coin two is $\frac{1}{2}$. To find the probability of flipping a tail with coin one and flipping a head with coin two, you must multiply the two probabilities together, $\frac{1}{2} \times \frac{1}{2} = \frac{1}{4}$. The probability of flipping a head with coin one and a tail with coin two is $\frac{1}{2} \times \frac{1}{2} = \frac{1}{4}$. Since it does not matter which coin is tails and which is heads, add the two probabilities together.

$$\frac{1}{4} + \frac{1}{4} = \frac{1}{2}$$

The theoretical probability is 50%.

Step 2: Find the experimental probability. The frequency of TH is 35, so out of 100 flips the probability is $\frac{35}{100}$. The frequency of HT is 23, so out of 100 flips, the probability is $\frac{23}{100}$. To find the theoretical probability of flipping one head and one tail, you need to add the two probabilities together.

$$\frac{35}{100} + \frac{23}{100} = \frac{58}{100} = \frac{29}{50}$$

The experimental probability based on the simulation is 58%.

Step 3: The difference between the theoretical probability and the experimental probability is 8%. Eight percent is not a huge difference. Since the two values are not too far apart, this means that the computer accurately simulates tossing two coins.

Use the simulations to find your answers.

1. A computer program simulated tossing three coins 500 times. The results are shown below.

HHH	50	HTT	66
HTH	76	THT	57
HHT	62	TTH	69
THH	64	TTT	56

(A) Based on the computer simulation, what is the experimental probability of tossing two heads and a tail?

(B) What is the theoretical probability of tossing two heads and one tail?

(C) Based on the computer simulation, what is the experimental probability of tossing three tails?

(D) What is the theoretical probability of tossing three tails?

(E) Compare your answers from part a with part b and compare your answer from part c with part d. Based on this comparison, is this an accurate simulation of tossing three coins?

2. Below is a computer simulation of rolling one six-sided cube 50 times.

Outcome	1	2	3	4	5	6
Frequency	8	6	12	11	5	7

(A) What is the theoretical probability of rolling a 3 or a 4?

(B) Calculate the theoretical probability of rolling a 6.

(C) Determine what the experimental probability of rolling a six based on the simulation.

(D) Compare the theoretical and experimental probabilities of rolling a six from parts b and c, what are your conclusions?

Chapter 9 Review

1. There are 50 students in the school orchestra in the following sections:

    25   string section
    15   woodwind
     5   percussion
     5   brass

One student is chosen at random to present the orchestra director with an award. What is the probability the student is from the woodwind section?

2. Fluffy's cat treat box contains 6 chicken-flavored treats, 5 beef-flavored treats, and 7 fish-flavored treats. If Fluffy's owner reaches in the box without looking, and chooses one treat, what is the probability that Fluffy will get a chicken-flavored treat?

3. The spinner stops on the number 5 on the first spin. What is the probability that it will not stop on 5 on the second spin?

4. Sherri turns the spinner in 3 times. What is the probability that the pointer always lands on a shaded number?

5. Three cakes are sliced into 20 pieces each. Each cake contains 1 gold ring. What is the probability that one person who eats one piece of cake from each of the 3 cakes will find 3 gold rings?

6. Brianna tosses a coin 4 times. What is the probability she gets all tails?

7. A box of a dozen doughnuts has 3 lemon cream-filled, 5 chocolate cream-filled, and 4 vanilla cream-filled. If the doughnuts look identical, what is the probability that if you pick a doughnut at random, it will be chocolate cream-filled?

8. Erica gets a her own cell phone. What is the probability that the last four digits of the phone are all 5's?

9. There are 26 letters in the alphabet. What is the probability that the first two letters of your new license plate will be your initials?

10. Mary has 4 green mints and 8 white mints of the same size in her pocket. If she picks out one, what is the probability it will be green?

Read the following, and answer questions 11–14.

There are 9 slips of paper in a hat, each with a number from 1 to 9. The numbers correspond to a group of students who must answer a question when the number for their group is drawn. Each time a number is drawn, the number is put back in the hat.

11. What is the probability that the number 6 will be drawn twice in a row?

12. What is the probability that the first 5 numbers drawn will be odd numbers?

13. What is the probability that the second, third, and fourth numbers drawn will be even numbers?

14. What is the probability that the first five times a number is drawn it will be the number 5?

Chapter 10
Measurement

This chapter covers the following Georgia Performance Standards:

M6M	Measurement	M6M1
		M6M2.a
M6P	Process Skills	M6P1.b
		M6P4.c

10.1   Using the Ruler

Practice measuring the object below with a ruler.

Measure these distances in inches.

1. How tall is the calculator?

2. How wide is the maple leaf?

3. How long is the car?

4. How far is it from the nose of the airplane to the nose of the camel?

5. How long is the trumpet?

6. How far is it from the middle of the bicycle's back wheel to the middle of the front wheel?

7. How long is the hour hand on the clock?

8. How far is it from the nose of the car to the mouthpiece of the trumpet?

10.2   More Measuring

Measure the following line segments to the nearest fourth of an inch.

1. ▬▬▬▬▬▬▬   = _____ in

2. ▬▬▬▬   = _____ in

3. ▬▬▬▬▬   = _____ in

4. ▬▬▬▬▬▬   = _____ in

5. ▬▬▬▬   = _____ in

Measure the dimensions of the following figures to the nearest sixteenth of an inch.

6.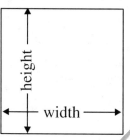

   height: _____ in
   width: _____ in

7.

   height from the top of the stem to the
   bottom: _____ in

8.

   width of the desk top: _____ in
   height of the desk: _____ in

9.

   length of guitar: _____ in

10.

   width of soccer ball: _____ in

11.

   height of tent: _____ in
   width of tent at its base: _____ in

10.3   Customary Measure

Customary measure in the United States is based on the English system. The following chart gives common customary units of measure as well as the standard units for time.

English System of Measure

Measure	Abbreviations	Appropriate Instrument
Time: 1 week = 7 days 1 day = 24 hours 1 hour = 60 minutes 1 minute = 60 seconds	week = wk hour = hr or h minutes = min seconds = sec	calendar clock clock clock
Length: 1 mile = 5,280 feet 1 yard = 3 feet 1 foot = 12 inches	mile = mi yard = yd foot = ft inch = in	odometer yard stick, tape line ruler, yard stick
Volume: 1 gallon = 4 quarts 1 quart = 2 pints 1 pint = 2 cups 1 cup = 8 ounces	gallon = gal quart = qt pint = pt ounce = oz	quart or gallon container quart container cup, pint, or quart container cup
Weight: 1 pound = 16 ounces	pound = lb ounce = oz	scale or balance
Temperature: Fahrenheit Celsius	°F °C	thermometer thermometer

10.4   Approximate English Measure

Match the item on the left with its approximate (not exact) measure on the right. You may use some answers more than once.

1. The height of an average woman is about _____.
2. An average candy bar weighs about _____.
3. An average doughnut is about _____ across (in diameter).
4. A piece of notebook paper is about _____ long.
5. A tennis ball is about _____ across (in diameter).
6. The average basketball is about _____ across.
8. How long is the average lunch table?
9. About how much does a computer disk weigh?
10. What is the average height of a table?

A. 1 yard
B. 2 yards
C. $5\frac{1}{2}$ feet
D. 4 weeks
E. $2\frac{1}{2}$ inches
F. 2 ounces
G. 1 foot

10.5   The Metric System

The metric system uses units based on multiples of ten. The basic units based on multiples of ten. The basic units of measure in the metric system are the meter, the liter, and the gram. Metric prefixes tell what multiple of ten the basic unit is multiplied by. Below is a chart of metric prefixes and their values. The ones rarely used are shaded.

Prefix	kilo (k)	hecto (h)	deka (da)	unit (m, L, g)	deci (d)	centi (c)	milli (m)
Meaning	1000	100	10	1	0.1	0.01	0.001

Multiply when changing from a greater unit to a smaller one; **divide** when changing from a smaller unit to a larger one. **The chart is set up to help you know how far and which direction to move a decimal point when making conversions from one unit to another.**

10.6   Understanding Meters

The basic unit of **length** in the metric system is the **meter**. Meter is abbreviated "m".

Metric Unit	Abbreviation	Memory Tip	Equivalents
1 millimeter	mm	Thickness of a dime	10 mm = 1 cm
1 centimeter	cm	Width of the tip of the little finger	100 cm = 1 m
1 meter	m	Distance from the nose to the tip of fingers (a little longer than a yard)	1000 m = 1 km
1 kilometer	km	A little more than half a mile	

10.7   Understanding Liters

The basic unit of **liquid volume** in the metric system is the **liter**. Liter is abbreviated "L".

The liter is the volume of a cube measuring 10 cm on each side. A milliliter is the volume of a cube measuring 1 cm on each side. A capital L is used to signify liter, so it is not confused with the number 1.

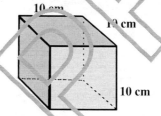

Volume = 1000 cm³ = 1 liter
(a little more than a quart)

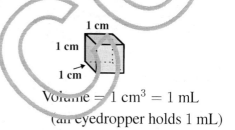

Volume = 1 cm³ = 1 mL
(an eyedropper holds 1 mL)

10.8   Understanding Grams

The basic unit of **mass** in the metric system is the **gram**. Gram is abbreviated "g".

A large paper clip weighs about 1 gram (1 g).

1000 grams = 1 kilogram (kg) = a little over 2 pounds

1 milligram (mg) = 0.001 gram. This is an extremely small amount and is used in medicine. An aspirin tablet weighs 300 mg.

10.9 Estimating Metric Measurements

Choose the best estimates.

1. The height of an average man
   (A) 18 cm
   (B) 1.8 m
   (C) 6 km
   (D) 36 mm

2. The volume of a coffee cup
   (A) 300 mL
   (B) 20 L
   (C) 5 L
   (D) 1 kL

3. The width of this book
   (A) 215 mm
   (B) 75 cm
   (C) 2 m
   (D) 1.5 km

4. The weight of an average man
   (A) 5 mg
   (B) 15 cg
   (C) 25 g
   (D) 90 kg

5. The length of a basketball player's foot
   (A) 2 m
   (B) 1 km
   (C) 30 cm
   (D) 100 mm

6. The weight of a dime
   (A) 3 g
   (B) 30 g
   (C) 10 cg
   (D) 1 kg

7. The width of your hand
   (A) 2 km
   (B) 0.5 m
   (C) 25 cm
   (D) 90 mm

8. The length of a basketball court
   (A) 1000 mm
   (B) 250 cm
   (C) 28 m
   (D) 2 km

Choose the best unit of measure.

9. The distance from Baton Rouge to Shreveport
   (A) millimeter
   (B) centimeter
   (C) meter
   (D) kilometer

10. The length of a house key
    (A) millimeter
    (B) centimeter
    (C) meter
    (D) kilometer

11. The thickness of a nickel
    (A) millimeter
    (B) centimeter
    (C) meter
    (D) kilometer

12. The width of a classroom
    (A) millimeter
    (B) centimeter
    (C) meter
    (D) kilometer

13. The length of a piece of chalk
    (A) millimeter
    (B) centimeter
    (C) meter
    (D) kilometer

14. The height of a pine tree
    (A) millimeter
    (B) centimeter
    (C) meter
    (D) kilometer

10.10   Converting Units within the Metric System

Converting units such as kilograms to grams or centimeters to decimeters is easy now that you know how to multiply and divide by multiples of ten.

Prefix	kilo (k)	hecto (h)	deka (da)	unit (m, L, g)	deci (d)	centi (c)	milli (m)
Meaning	1000	100	10	1	0.1	0.01	0.001

Example 1:   2 L =____mL

2.000 L = 2000 mL

Look at the chart above. To move from liters to milliliters, you move to the right three places. So, to convert the 2 L to mL, move the decimal point three places to the right. You will need to add two zeros.

Example 2:   5.25 cm =____m

005.25 cm = 0.0525 m

To move from centimeters to meters, you need to move two spaces to the left. So, to convert 5.25 cm to m, move the decimal point two spaces to the left. Again, you need to add zeros.

Solve the following problems.

1.   35 mg = _____ g

2.   6 km = _____ m

3.   21.5 mL = _____ L

4.   4.9 mm = _____ cm

5.   5.35 kL = _____ mL

6.   32.1 mg = _____ kg

7.   156.4 m = _____ km

8.   25 mg = _____ cg

9.   17.5 L = _____ mL

10.   4.2 g = _____ kg

11.   0.06 daL = _____ dL

12.   0.417 kg = _____ cg

13.   18.2 cL = _____ L

14.   81.2 dm = _____ cm

15.   72.3 cm = _____ m

16.   0.003 kL = _____ L

17.   5.06 g = _____ mg

18.   1.058 mL = _____ cL

19.   43 hm = _____ km

20.   2.057 m = _____ cm

21.   564.3 g = _____ kg

Chapter 10 Review

Fill in the blanks below with the appropriate unit of measurement.

1. A box of assorted chocolates might weigh about 1 _____ (English).

2. A compact disc is about 7 _____ (English) across.

3. In Europe, gasoline is sold in _____ (metric).

4. A vitamin C tablet has a mass of 500 _____(metric).

Fill in the blanks below with the appropriate English or metric conversions.

5. Two gallons equals _____ cups.

6. 4.2 L equals _____ mL.

7. 31 yards equals _____ inches.

8. 6,800 m equals _____ kilometers.

9. 36 oz. equals _____ pounds.

10. 730 mg equals _____ kg.

Solve the following problems.

11.   120 m = _____ km   14.   1.5 mg = _____ g   17.   0.005 kg = _____ g

12.   9 g = _____ mg   15.   15 cm = _____ mm   18.   55 mL = _____ L

13.   0.02 kL = _____ L   16.   5 L = _____ mL   19.   30 cm = _____ m

Chapter 11
Plane Geometry

This chapter covers the following Georgia Performance Standards:

M6M	Measurement	M6M2.b, c
M6G	Geometry	M6G1.c
M6P	Process Skills	M6P1.b
		M6P3.d
		M6P4.c
		M6P5.a, b, c

11.1 Perimeter

The **perimeter** is the distance around a polygon. To find the perimeter, add the lengths of the sides.

Examples:

$P = 7 + 15 + 7 + 15$
$P = 44 \text{ in}$

$P = 4 + 6 + 5$
$P = 15 \text{ cm}$

$P = 8 + 15 + 20 + 12 + 10$
$P = 65 \text{ ft}$

Find the perimeter of the following polygons.

1.
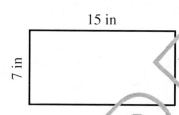

3.
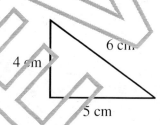

5.
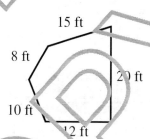

2.
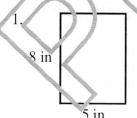

4.
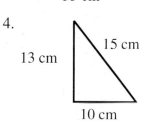

6.
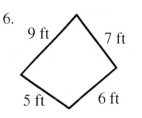

11.2   Area of Squares and Rectangles

Area - area is always expressed in square units, such as in^2, m^2, and ft^2.

The area, (A), of squares and rectangles equals length (l) times width (w). $A = l \times w$.

Example 1:

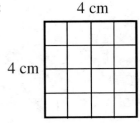

4 cm

4 cm

$A = lw$
$A = 4 \times 4$
$A = 16 \text{ cm}^2$

If a square has an area of 16 cm^2, it means that it will take 16 squares that are 1 cm on each side to cover the area that is 4 cm on each side.

Find the area of the following squares and rectangles using the formula $A = lw$.

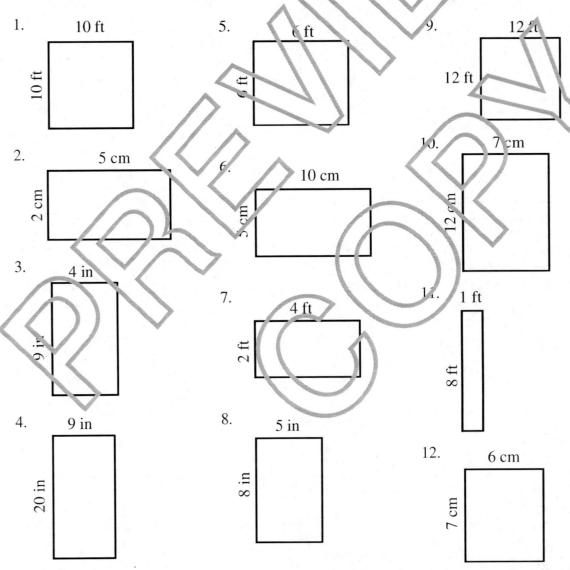

1.   10 ft
     10 ft

2.   5 cm
     2 cm

3.   4 in
     9 in

4.   9 in
     20 in

5.   6 ft
     6 ft

6.   10 cm
     5 cm

7.   4 ft
     2 ft

8.   5 in
     8 in

9.   12 ft
     12 ft

10. 7 cm
     12 cm

11. 1 ft
     8 ft

12. 6 cm
     7 cm

11.3   Area of Triangles

Example 2:   Find the area of the following triangle.

The formula for the area of a triangle is as follows:

$$A = \frac{1}{2} \times b \times h$$

A = area
b = base
h = height or altitude

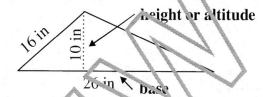

Step 1:   Insert the measurements from the triangle into the formula: $A = \frac{1}{2} \times 26 \times 10$

Step 2:   Cancel and multiply. $A = \frac{1}{2} \times \frac{\overset{13}{26}}{1} \times \frac{10}{1} = 130 \text{ in}^2$

Note: Area is always expressed in square units such as in^2, ft^2, or m^2.

Find the area of the following triangles. Remember to include units.

1.

3 in, 5 in, 4 in

5.

3 ft, 2 ft

9.

2 ft

2.

7 cm, 6 cm, 12 cm, height

6.

20 cm

10.

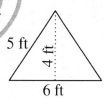
5 ft, 4 ft, 6 ft

3.

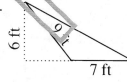
6 ft, 9 in, 7 ft

7.

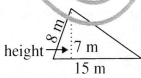
8 m, height, 7 m, 15 m

11.

12 ft, 10 ft, 15 ft

4.

12 cm, 12 cm

8.

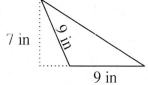
7 in, 9 in, 9 in

12.

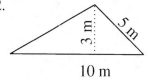
3 m, 5 m, 10 m

11.4   Circumference

Circumference, C - the distance around the outside of a circle

Diameter, d - a line segment passing through the center of a circle from one side to the other

Radius, r - a line segment from the center of a circle to the edge of a circle

Pi, π - the ratio of a circumference of a circle to its diameter $\pi = 3.14$ or $\pi = \dfrac{22}{7}$

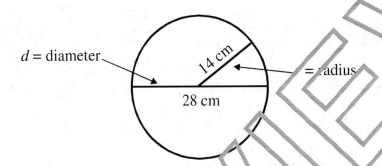

d = diameter

14 cm

28 cm

= radius

The formula for the circumference of a circle is $C = 2\pi r$ or $C = \pi d$. (The formulas are equal because the diameter is equal to twice the radius, $d = 2r$.)

Example 3:   Find the circumference of the circle above.

$C = \pi d$ Use $\pi = 3.14$ $C = 2\pi r$
$C = 3.14 \times 28$ $C = 2 \times 3.14 \times 14$
$C = 87.92$ cm $C = 87.92$ cm

Use the formulas given above to find the circumferences of the following circles. Use $\pi = 3.14$.

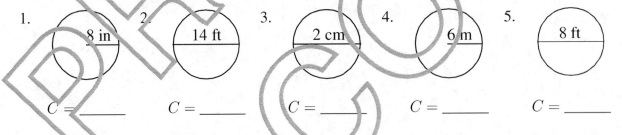

1. 8 in   2. 14 ft   3. 2 cm   4. 6 m   5. 8 ft

$C =$ _____ $C =$ _____ $C =$ _____ $C =$ _____ $C =$ _____

Use the formulas given above to find the circumferences of the following circles. Use $\pi = \dfrac{22}{7}$.

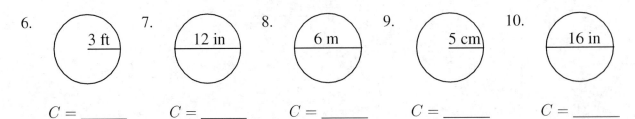

6. 3 ft   7. 12 in   8. 6 m   9. 5 cm   10. 16 in

$C =$ _____ $C =$ _____ $C =$ _____ $C =$ _____ $C =$ _____

11.5   Area of a Circle

The formula for the area of a circle is $A = \pi r^2$. The area is how many square units of measure would fit inside a circle.

Example 4:   Find the area of the circle, using both values for π.

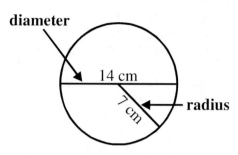

diameter

14 cm

7 cm — radius

Let $\pi = \dfrac{22}{7}$

$A = \pi r^2$

$A = \dfrac{22}{7} \times 7^2$

$A = \dfrac{22}{7} \times \dfrac{49}{1} \, \dfrac{7}{1}$

$= 154 \text{ cm}^2$

Let $\pi = 3.14$

$A = \pi r^2$

$A = 3.14 \times 7^2$

$A = 3.14 \times 49$

$= 153.86 \text{ cm}^2$

Find the area of the following circles. Remember to include units.

$\pi = 3.14 \qquad \pi = \dfrac{22}{7}$

1.
   5 in

   $A = \_\_\_\_ \qquad A = \_\_\_\_$

2.
   16 ft

   $A = \_\_\_\_ \qquad A = \_\_\_\_$

3.
   8 cm

   $A = \_\_\_\_ \qquad A = \_\_\_\_$

4.
   3 in

   $A = \_\_\_\_ \qquad A = \_\_\_\_$

Fill in the chart below. Include appropriate units.

		Area	
Radius	Diameter	$\pi = 3.14$	$\pi = \dfrac{22}{7}$
5. 9 ft			
6.	4 in		
7. 8 cm			
8.	20 ft		
9. 14 in			
10.	18 cm		
11. 12 ft			
12.	6 in		

11.6   Two-Step Area Problems

Solving the following problems will require two steps. You will need to find the area of two figures, and then either add or subtract the two areas to find the answer. **Carefully read Examples 5 and 6 on the next page.**

Example 5:	**Example 6:**
Find the area of the living room below.	**Find the area of the shaded sidewalk.**

Figure 1

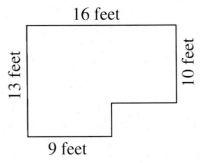

Step 1: Complete the rectangle as in Figure 2, and compute the area as if it were a complete rectangle.

Figure 2

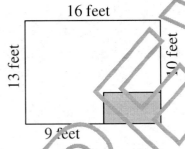

$$A = \text{length} \times \text{width}$$
$$A = 16 \times 13$$
$$A = 208 \ \text{ft}^2$$

Step 2: Figure the area of the shaded part.

7 feet

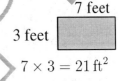

3 feet

$$7 \times 3 = 21 \ \text{ft}^2$$

Step 3: Subtract the area of the shaded part from the area of the complete rectangle

$$208 - 21 = 187 \ \text{ft}^2$$

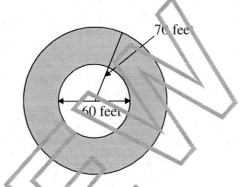

Step 1: Find the area of the outside circle.
$$\pi = 3.14$$
$$A = 3.14 \times 70 \times 70$$
$$A = 15,386 \ \text{ft}^2$$

Step 2: Find the area of the inside circle.
$$\pi = 3.14$$
$$A = 3.14 \times 30 \times 30$$
$$A = 2826 \ \text{ft}^2$$

Step 3: Subtract the area of the inside circle from the area of the outside circle.
$$15,386 - 2826 = 12,560 \ \text{ft}^2$$

Find the area of the following figures.

1.

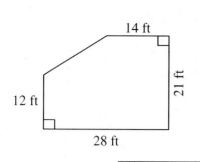

2.

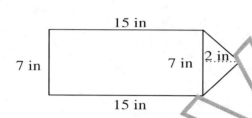

3. What is the area of the shaded circle?
    Use $\pi = 3.14$, and round the answer to
    the nearest whole number.

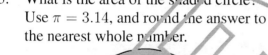

4.

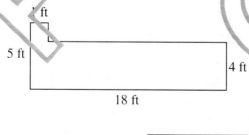

5.

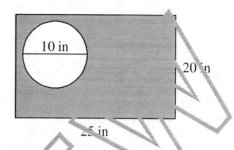

6. What is the area of the shaded part?

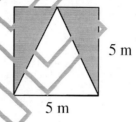

7. What is the area of the shaded part?

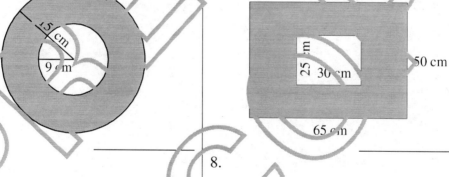

8.

11.7   Similar and Congruent

Similar figures have the same shape but are two different sizes. Their corresponding sides are proportional. **Congruent figures** are exactly alike in size and shape and their corresponding sides are equal. See the examples below.

SIMILAR   **CONGRUENT**

Label each pair of figures below as either S if they are similar, C if they are congruent, or N if they are neither.

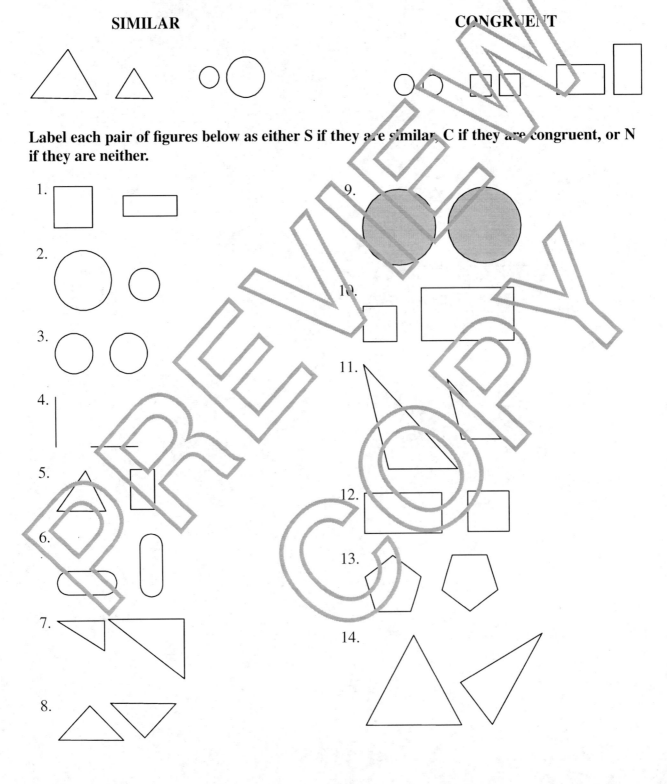

Copyright © American Book Company

11.8   Similar Triangles

Two triangles are similar if the measurements of the three angles in both triangles are the same. If the three angles are the same, then their corresponding sides are proportional. The similar triangles are proportional by a **scale factor**, k. This is expressed in the equation $y = xk$.

Corresponding Sides - The triangles below are similar. Therefore, the two shortest sides from each triangle, c and f, are corresponding. The two longest sides from each triangle, a and d, are corresponding. The two medium length sides, b and e, are corresponding.

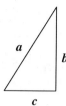 

Proportional - The corresponding sides of similar triangles are proportional to each other. This means if we know all the measurements of one triangle, and we only know one measurement of the other triangle, we can figure out the measurements of the two other sides with proportion problems. The two triangles below are similar.

 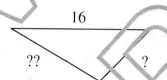

Note: To set up the proportion correctly, it is important to keep the measurements of each triangle on opposite sides of the equal sign.

To find the short side:

Step 1: Set up the proportion

$$\frac{\text{long side}}{\text{short side}} \quad \frac{12}{6} = \frac{16}{?}$$

Step 2: Solve the proportion. Multiply the two numbers diagonal to each other and then divide by the other number.
$$16 \times 6 = 96$$
$$96 \div 12 = 8$$

To find the medium length side:

Step 1: Set up the proportion

$$\frac{\text{long side}}{\text{medium}} \quad \frac{12}{9} = \frac{16}{??}$$

Step 2: Solve the proportion. Multiply the two numbers diagonal to each other and then divide by the other number.
$$16 \times 9 = 144$$
$$144 \div 12 = 12$$

To find the scale factor in the problem on the previous page, we must divide the a value from the second triangle by the corresponding value from the first triangle. The value 16 is from the second triangle, and the corresponding value from the first triangle is 12. $k = \dfrac{16}{12} = \dfrac{4}{3}$

The scale factor in this problem is $\frac{4}{3}$.

To check this answer multiply every term in the first triangle by the scale factor, and you will get every term in the second triangle.

$$12 \times \frac{4}{3} = 16 \qquad 9 \times \frac{4}{3} = 12 \qquad 6 \times \frac{4}{3} = 8$$

Find the missing side from the following similar triangles

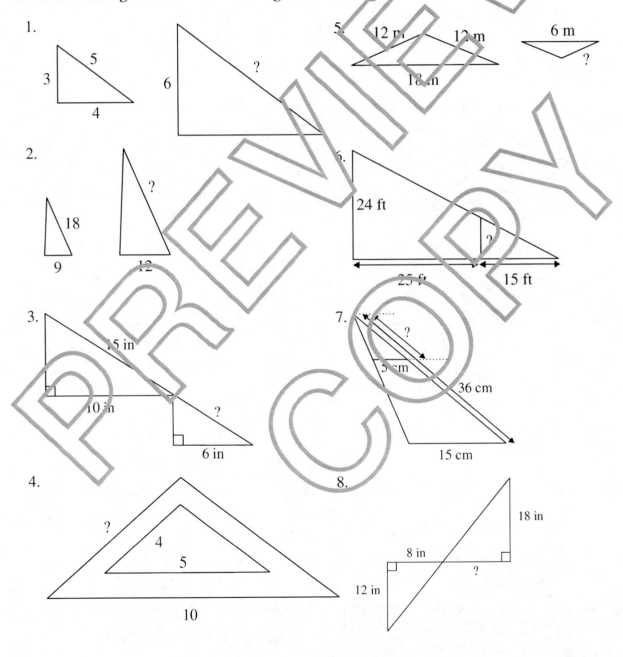

1.

2.

3.

4.

5.

6.

7.

8.

Chapter 11 Review

Chapter 11 Review

1. What is the length of the line segment $\overline{WY}$?

_____

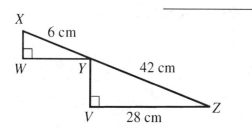

2. Find the area of the shaded region of the figure below.

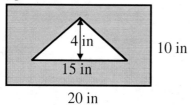

$A =$ _____

3. Calculate the perimeter of the following figure.

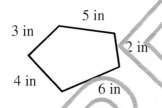

$P =$ _____

4. Calculate the perimeter and area of the following figure.

$P =$ _____
$A =$ _____

5. Calculate the circumference and the area of the following circle. Use $\pi = 3.14$.

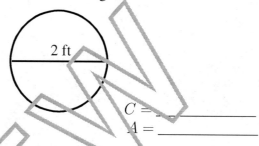

$C =$ _____
$A =$ _____

6. Find the area of the shaded part.

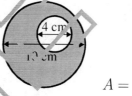

$A =$ _____

7. The following two triangles are similar. Find the length of the missing side.

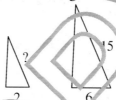

_____

8. Find the missing side of the triangle below.

_____

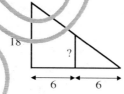

Copyright © American Book Company 113

Chapter 12
Solid Geometry

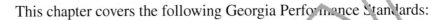

This chapter covers the following Georgia Performance Standards:

M6M	Measurement	M6M2.b, c
		M6M3.a, b, c, d
		M6M4.a, b, c, d
M6G	Geometry	M6G2.a, b, c, d
M6P	Process Skills	M6P1.b
		M6P3.d
		M6P4.b, c
		M6P5.a, b, c

12.1   Understanding Volume

Volume - Measurement of volume is expressed in cubic units such as in^3, ft^3, m^3, cm^3, or mm^3. The volume of a solid is the number of cubic units that can be contained in the solid.

First, let's look at rectangular solids.

Example 1:   How many 1 cubic centimeter cubes will it take to fill up the figure below?

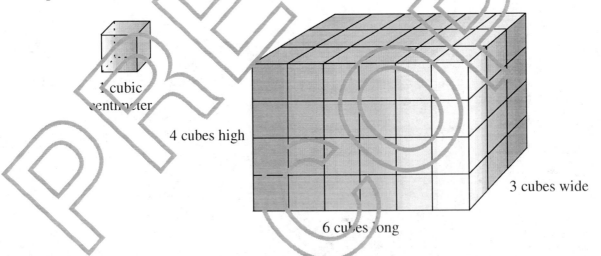

To find the volume, you need to multiply the length times the width times the height.

Volume of a rectangular solid = length $\times$ width $\times$ height $(V = lwh)$.

$$V = 6 \times 3 \times 4 = 72\,\text{cm}^3$$

12.2   Volume of Rectangular Prisms

You can calculate the volume (V) of a rectangular prism (box) by multiplying the length (l) by the width (w) by the height (h), as expressed in the formula $V = (lwh)$.

Example 2:   Find the volume of the box pictured here:

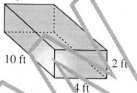

Step 1:   Insert measurements from the figure into the formula.

Step 2:   Multiply to solve. $10 \times 4 \times 2 = 80 \text{ ft}^3$

Note: **Volume is always expressed in cubic units such as in^3, ft^3, m^3, cm^3, or mm^3.**

Find the volume of the following rectangular prisms (boxes).

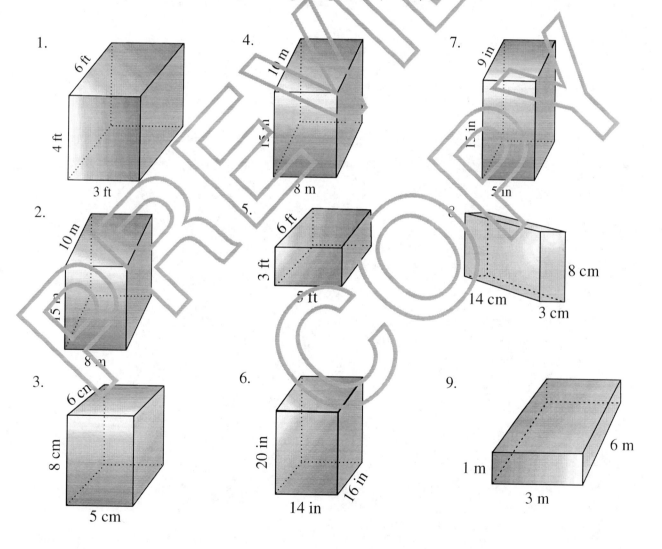

12.3   Volume of Cubes

A **cube** is a special kind of rectangular prism (box). Each side of a cube has the same measure. So, the formula for the volume of a cube is $V = s^3$ ($s \times s \times s$).

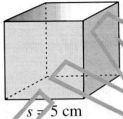

Example 3:   Find the volume of the cube at right:

$s = 5$ cm

Step 1:   Insert measurements from the figure into the formula.

Step 2:   Multiply to solve. $5 \times 5 \times 5 = 125$ cm^3

Note: Volume is always expressed in cubic units such as in^3, ft^3, m^3, cm^3, or mm^3.

Answer each of the following questions about cubes.

1. If a cube is 3 centimeters on each edge, what is the volume of the cube?

Find the volume of the following cubes.

7.

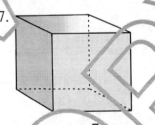

$s = 7$ in.

2. If the measure of the edge is doubled to 6 centimeters on each edge, what is the volume of the cube?

8.

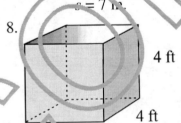

4 ft

3. What if the edge of a 3 centimeter cube is tripled to become 9 centimeters on each edge? What will the volume be?

4 ft

4 ft

4. How many cubes with edges measuring 3 centimeters would you need to stack together to make a solid 12 centimeter cube?

9. 12 inches = 1 foot

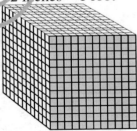

5. What is the volume of a 2-centimeter cube?

6. Jerry built a 2-inch cube to hold his marble collection. He wants to build a cube with a volume 8 times larger. How much will each edge measure?

$s = 1$ foot
How many cubic inches are in a cubic foot?

12.4 Volume of Spheres, Cones, Cylinders, and Pyramids

To find the volume of a solid, insert the measurements given for the solid into the correct formula and solve. Remember, volumes are expressed in cubic units such as in^3, ft^3, m^3, cm^3, or mm^3.

Sphere	**Cone**	**Cylinder**
$V = \frac{4}{3}\pi r^3$	$V = \frac{1}{3}\pi r^2 h$	$V = \pi r^2 h$

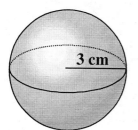 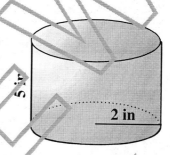

$V = \frac{4}{3}\pi r^3 \quad \pi = 3.14$

$V = \frac{4}{3} \times 3.14 \times 27$

$V = 113.04 \, cm^3$

$V = \frac{1}{3}\pi r^2 h \quad \pi = 3.14$

$V = \frac{1}{3} \times 3.14 \times 49 \times 10$

$V = 512.87 \, in^3$

$V = \pi r^2 h \quad \pi = \frac{22}{7}$

$V = \frac{22}{7} \times 4 \times 5$

$V = 62\frac{6}{7} \, in^3$

Pyramids

$V = \frac{1}{3}Bh \quad B = $ area of rectangular base

$V = \frac{1}{3}Bh \quad B = $ area of triangular base

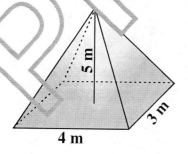 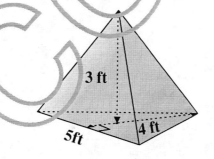

$V = \frac{1}{3}Bh \quad B = l \times w$

$V = \frac{1}{3} \times 4 \times 3 \times 5$

$V = 20 \, m^3$

$B = \frac{1}{2} \times 5 \times 4 = 10 \, ft^2$

$V = \frac{1}{3} \times 10 \times 3$

$V = 10 \, ft^3$

Find the volume of the following shapes. Use $\pi = 3.14$.

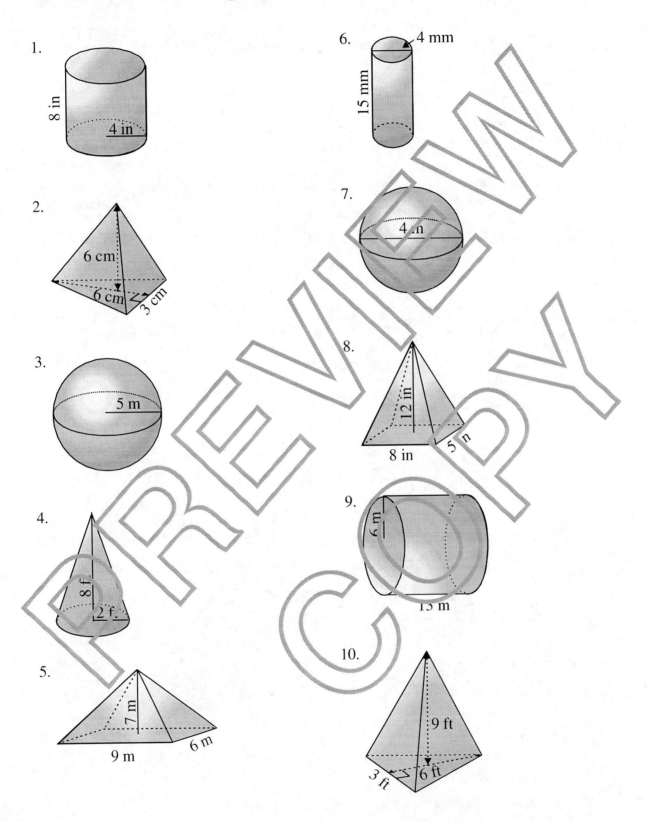

1.

8 in

4 in

2.

6 cm

6 cm

3 cm

3.

5 m

4.

8 f

2 f

5.

7 m

9 m

6 m

6.

4 mm

15 mm

7.

4 in

8.

12 in

8 in

5 n

9.

6 m

13 m

10.

9 ft

3 ft

6 ft

12.5 Two-Step Volume Problems

Some objects are made from two geometric figures. For example, the tower below is made up of two geometric objects.

Example 4: Find the maximum volume of the tower.

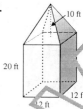

Step 1: Determine which formulas you will need. The tower is made from a pyramid and a rectangular prism, so you will need the formulas for the volume of these two figures.

Step 2: Find the volume of each part of the tower. The bottom of the tower is a rectangular prism $V = lwh$
$V = 12 \times 12 \times 20 = 2,880 \text{ ft}^3$
The top of the tower is a rectangular pyramid $V = \frac{1}{3}Bh$
$V = \frac{1}{3} \times 12 \times 12 \times 10 = 480 \text{ ft}^3$

Step 3: Add the two volumes together. $2800 \text{ ft}^3 + 480 \text{ ft}^3 = 3,360 \text{ ft}^3$

Find the volume of the geometric figures below. Hint: If part of a solid has been removed, find the volume of the hole, and subtract it from the volume of the total object.

1.

2. Each side measures 3 inches.

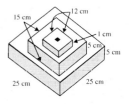

3. A rectangular hole passes through the middle of the figure below. The hole measure 1 cm on each side.

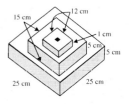

4. In the figure below, 3 cylinders are stacked on top of each other. The radii of the cylinders are 2 inches, 4 inches, and 6 inches. The height of each cylinder is 1 inch.

5.

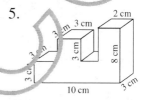

6. A hole, 1 meter in diameter, has been cut through the cylinder below.

12.6   Geometric Relationships of Solids

In the previous chapter, you looked at geometric relationships between 2-dimensional figures. Now you will learn about the relationships among 3-dimensional figures. The formulas for finding the volumes of geometric solids are given below.

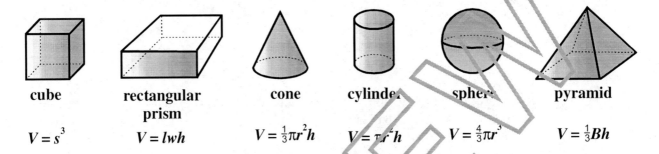

cube	rectangular prism	cone	cylinder	sphere	pyramid
$V = s^3$	$V = lwh$	$V = \frac{1}{3}\pi r^2 h$	$V = \pi r^2 h$	$V = \frac{4}{3}\pi r^3$	$V = \frac{1}{3}Bh$

By studying each formula and by comparing formulas between different solids, you can determine general relationships.

Example 5:   How would doubling the radius of a sphere affect the volume?

The volume of a sphere is $v = \frac{4}{3}\pi r^3$. Just by looking at the formula, can you see that by doubling the radius, the volume would increase 8 times the original volume? So, a sphere with a radius of 2 would have a volume 8 times greater than a sphere with a radius of 1.

Example 6:   A cylinder and a cone have the same radius and the same height. What is the difference between their volumes?

Compare the formulas for the volume of a cone and the volume of a cylinder. They are identical except that the cone is multiplied by $\frac{1}{3}$. Therefore, the volume of a cone with the same height and radius as a cylinder would be one-third less. Or, the volume of a cylinder with the same height and radius as a cone would be three times greater.

Example 7:   If you double one dimension of a rectangular prism, how will the volume be affected? How about doubling two dimensions? How about doubling all three dimensions?

Do you see that doubling just one of the dimensions of a rectangular prism will also double the volume? Doubling two of the dimensions will cause the volume to increase 4 times the original volume. Doubling all three dimensions will cause the volume to increase 8 times the original volume.

Example 8:   A cylinder holds 100 cubic centimeters of water. If you triple the radius of the cylinder but keep the height the same, how much water would you need to fill the new cylinder?

Tripling the radius of a cylinder causes the volume to increase by 3^2 or 9 times the original volume. The volume of the new cylinder would hold 9×100 or 900 cubic centimeters of water.

Answer the following questions by comparing the volumes of two solids that share some of the same dimensions.

1. If you have a cylinder with a height of 8 inches and a radius of 4 inches, and you have a cone with the same height and radius, how many times greater is the volume of the cylinder than the volume of the cone?

2.

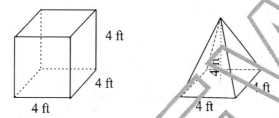

   In the two figures above, how many times larger is the volume of the cube than the volume of the pyramid?

3. How many times greater is the volume of a cylinder if you double the radius?

4. How many times greater is the volume of a cylinder if you double the height?

5. In a rectangular solid, how many times greater is the volume if you double the length?

6. In a rectangular solid, how many times greater is the volume if you double the length and the width?

7. In a rectangular solid, how many times greater is the volume if you double the length and the width and the height?

8. In the following two figures, how many cubes like Figure 1 will fit inside Figure 2?

9. A sphere has a radius of 1. If the radius is increased to 3, how many times greater will the volume be?

10. It takes 2 liters of water to fill cone A below. If the cone is stretched so the radius is doubled, but the height stays the same, how much water is needed to fill the new cone, B?

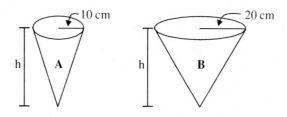

12.7   Surface Area

The **surface area of a solid** is the total area of all the sides of a solid.

12.8   Cube

There are six sides on a cube. To find the surface area of a cube, find the area of one side and multiply by 6.

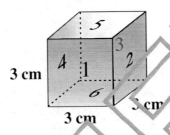

Area of each side of the cube: $3 \times 3 = 9 \text{ cm}^2$

Total surface area: $9 \times 6 = 54 \text{ cm}^2$

12.9   Rectangular Prism

There are 6 sides on a rectangular prism. To find the surface area, add the areas of the six rectangular sides.

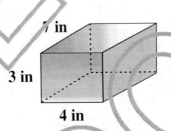

Top and Bottom

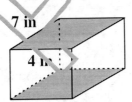

Area of top side:
7 in $\times$ 4 in $= 28 \text{ in}^2$
Area of top and bottom:
28 in $\times$ 2 $= 56 \text{ in}^2$

Front and Back

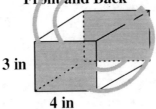

Area of front:
3 in $\times$ 4 in $= 12 \text{ in}^2$
Area of front and back:
12 in $\times$ 2 $= 24 \text{ in}^2$

Left and Right

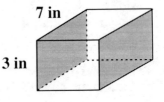

Area of left side:
3 in $\times$ 7 in $= 21 \text{ in}^2$
Area of left and right:
21 in $\times$ 2 $= 42 \text{ in}^2$

Total surface area: $56 \text{ in}^2 + 24 \text{ in}^2 + 42 \text{ in}^2 = 122 \text{ in}^2$

Find the surface area of the following cubes and prisms.

1.

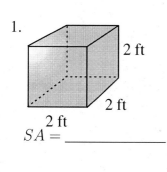

2 ft

2 ft

2 ft

2 ft

$SA =$ _____

2.

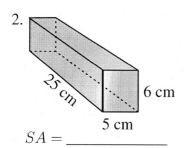

25 cm

6 cm

5 cm

$SA =$ _____

3.

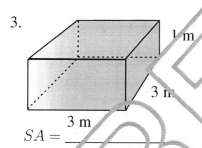

1 m

3 m

3 m

$SA =$ _____

4.

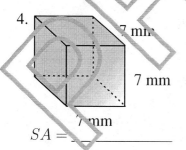

7 mm

7 mm

7 mm

$SA =$ _____

5.

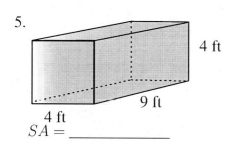

4 ft

9 ft

4 ft

$SA =$ _____

6.

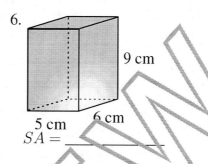

9 cm

5 cm   6 cm

$SA =$ _____

7.

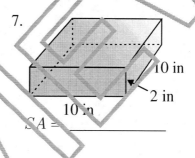

10 in

2 in

10 in

$SA =$ _____

8.

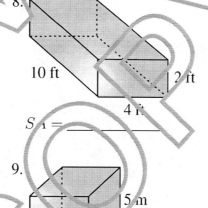

10 ft   2 ft

4 ft

$SA =$ _____

9.

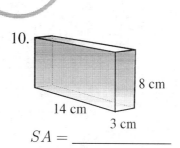

5 m

5 m

5 m

$SA =$ _____

10.

8 cm

14 cm

3 cm

$SA =$ _____

12.10   Pyramid

The pyramid below is made of a square base with 4 triangles on the sides.

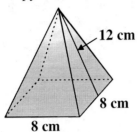

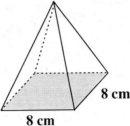

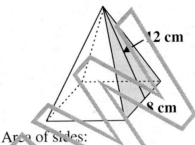

Area of square base:
$A = l \times w$
$A = 8 \times 8 = 64$ cm^2

Area of sides:
Area of 1 side $= \frac{1}{2}bh$
$A = \frac{1}{2} \times 8 \times 12 = 48$ cm^2
Area of 4 sides $= 48 \times 4 = 192$ cm^2

Total surface area: $64 + 192 = 256$ cm^2

Find the total surface of the following pyramids.

1.

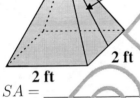

3 ft

2 ft

2 ft

$SA = \underline{\hspace{2cm}}$

4.

7 cm

8 cm   **8 cm**

$SA = \underline{\hspace{2cm}}$

7.

9 m

4 m   **4 m**

$SA = \underline{\hspace{2cm}}$

2.

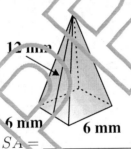

12 mm

6 mm   **6 mm**

$SA = \underline{\hspace{2cm}}$

5.

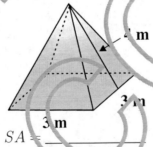

3 m

3 m

$SA = \underline{\hspace{2cm}}$

8.

10 in

5 in

5 in

$SA = \underline{\hspace{2cm}}$

3.

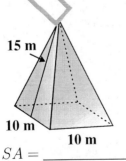

15 m

10 m
10 m

$SA = \underline{\hspace{2cm}}$

6.

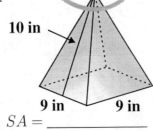

10 in

9 in   **9 in**

$SA = \underline{\hspace{2cm}}$

9.

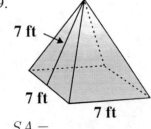

7 ft

7 ft
7 ft

$SA = \underline{\hspace{2cm}}$

12.11   Cylinder

If the side of a cylinder was slit from top to bottom and laid flat, its shape would be a rectangle. The length of the rectangle is the same as the circumference of the circle that is the base of the cylinder. The width of the rectangle is the height of the cylinder.

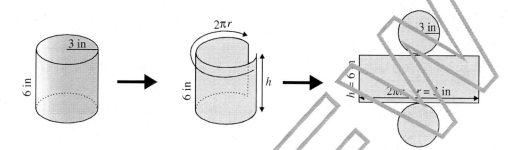

Total Surface Area of a Cylinder $= 2\pi r^2 + 2\pi rh$

Area of top and bottom:
Area of a circle $= \pi r^2$
Area of top $= 3.14 \times 3^2 = 28.26$ in^2
Area of top and bottom $= 2 \times 28.26 = 56.52$ in^2

Area of side:
Area of rectangle $= l \times h$
$l = 2\pi r = 2 \times 3.14 \times 3 = 18.84$ in
Area of rectangle $= 18.84 \times 6 = 113.04$ in^2

Total surface area $= 56.52 + 113.04 = 169.56$ in^2

Find the total surface area of the following cylinders. Use $\pi = 3.14$.

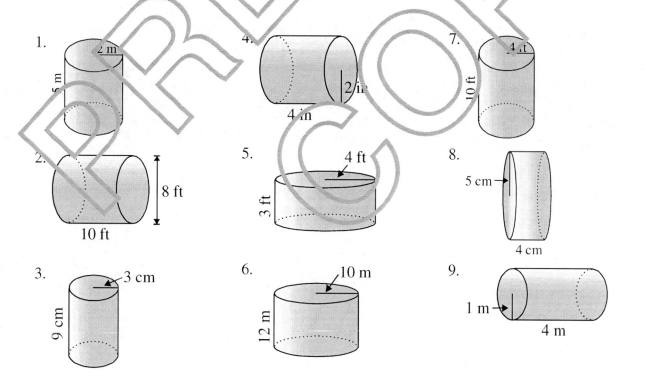

12.12   Sphere

Surface area $= 4\pi r^2$
Surface area $= 4 \times 3.14 \times 4^2$
Surface area $= 200.96$ cm^2

Find the surface area of a sphere given the following measurements where r = radius and d = diameter. Use $\pi = 3.14$.

1. $r = 2$ in SA = _____
2. $r = 6$ m SA = _____
3. $r = \frac{3}{4}$ yd SA = _____
4. $d = 8$ cm SA = _____
5. $d = 50$ mm SA = _____
6. $r = \frac{1}{4}$ ft SA = _____

7. $d = 14$ cm SA = _____
8. $r = \frac{1}{5}$ km SA = _____
9. $d = 3$ in SA = _____
10. $d = \frac{2}{3}$ ft SA = _____
11. $r = 10$ mm SA = _____
12. $d = 5$ yd SA = _____

12.13   Cone

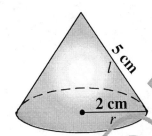

Total Surface Area: $T = \pi r (r + s)$
$\pi = 3.14$ $r =$ radius of base $s =$ slant height
$T = 3.14 \times 2 (2 + 5)$
$T = 6.28 \times 7$
$T = 43.96$ cm^2

Find the surface area of the following cones. Use $\pi = 3.14$.

1.

3.

5.

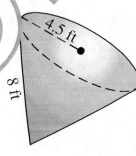

2.

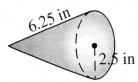

4.

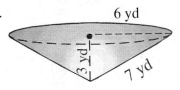

6.

12.14 Nets of Solid Objects

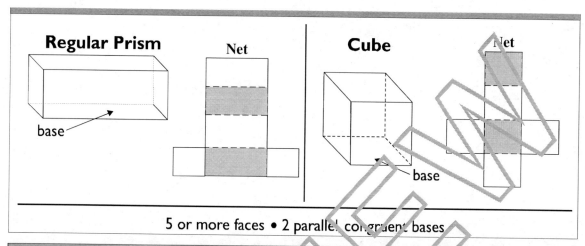

Regular Prism   Net

base

Cube   Net

base

5 or more faces • 2 parallel congruent bases

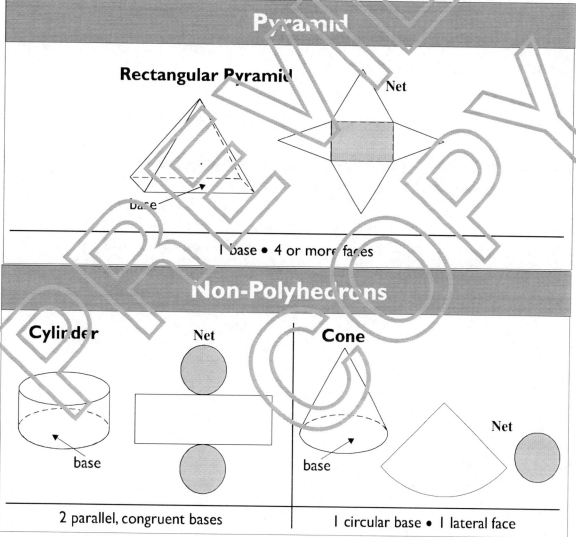

Pyramid

Rectangular Pyramid   Net

base

1 base • 4 or more faces

Non-Polyhedrons

Cylinder   Net

base

Cone   Net

base

2 parallel, congruent bases   1 circular base • 1 lateral face

12.15   Using Nets to Find Surface Area

A **net** is a two-dimensional representation of a three-dimensional object. Nets clearly illustrate the plane figures that make up a solid.

Example 9:   Find the surface area of the figure shown below.

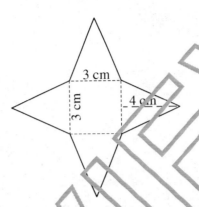

Step 1:   Find the area of the 4 triangles
$A = \frac{1}{2}bh = \frac{1}{2} \times 3 \times 4 = 6$ cm^2
Area of all 4 triangles $= 4 \times 6 = 24$ cm^2

Step 2:   Find the area of the base.
$A = lw = 3 \times 3 = 9$ cm^2

Step 3:   Find the sum of the areas of all the plane figures.
Surface Area $= 24$ cm$^2 + 9$ cm^2
$SA = 33$ cm^2

Example 10:   A net for a cone is shown below. Find the surface area of each part to find the total surface area.

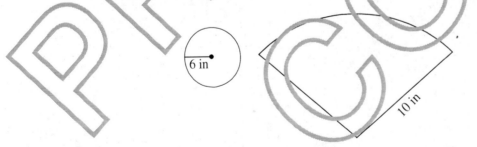

Step 1:   Find the area of the base.
$A = \pi r^2 = 3.14 \times 6^2 = 3.14 \times 36 = 113.04$ in^2

Step 2:   Find the area of the cone section.
$A = \pi r s = 3.14 \times 6 \times 10 = 188.40$ in^2

Step 3:   Find the sum of the areas of the base and the cone section.
Surface area $= 113.04$ in$^2 + 188.40$ in^2
$SA = 301.44$ in^2

The nets for the various solids are given. Find the surface area of the objects. If needed, use
$\pi = 3.14$.

1.

3 in

3 in

3.

5 cm

11 cm

2.

5 cm

8 cm   7 cm

4.

7 ft

15 ft

7 ft

Using a ruler, measure the dimensions of the following nets to the nearest tenth of a centimeter, and calculate the surface area of the object. If needed, use $\pi = 3.14$.

5.

6.

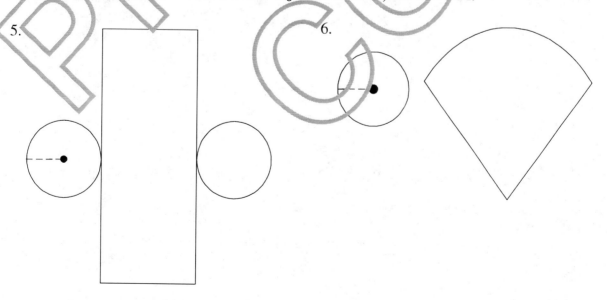

12.16   Solid Geometry Word Problems

1. If an Egyptian pyramid has a square base that measures 500 yards by 500 yards, and the pyramid stands 300 yards tall, what would be the volume of the pyramid? Use the formula for volume of a pyramid, $V = \frac{1}{3}Bh$ where B is the area of the base.

2. Robert is using a cylindrical barrel filled with water to flatten the soil in his yard. The circular ends have a radius of 1 foot. The barrel is 3 feet wide. How much water will the barrel hold? The formula for volume of a cylinder is $V = \pi r^2 h$. Use $\pi = 3.14$.

3. If a basketball measures 24 centimeters in diameter, what volume of air will it hold? The formula for volume of a sphere is $V = \frac{4}{3}\pi r^3$. Use $\pi = 3.14$.

4. What is the volume of a cone that is 2 inches in diameter and 5 inches tall? The formula for volume of a cone is $V = \frac{1}{3}\pi r^2 h$. Use $\pi = 3.14$.

5. Kelly has a rectangular fish aquarium that measures 24 inches wide, 12 inches deep, and 18 inches tall. What is the maximum amount of water that the aquarium will hold?

6. Jenny has a rectangular box that she wants to cover in decorative contact paper. The box is 10 cm long, 5 cm wide, and 5 cm high. How much paper will she need to cover all 6 sides?

7. Gasco needs to construct a cylindrical steel gas tank that measures 6 feet in diameter and is 8 feet long. How many square feet of steel will be needed to construct the tank? Use the following formulas as needed: $A = l \times w$, $A = \pi r^2$, $C = 2\pi r$. Use $\pi = 3.14$.

8. Craig wants to build a miniature replica of San Francisco's Transamerica Pyramid out of glass. His replica will have a square base that measures 6 cm by 6 cm. The 4 triangular sides will be 6 cm wide and 60 cm tall. How many square centimeters of glass will he need to build his replica? Use the following formulas as needed: $A = l \times w$ and $A = \frac{1}{2}bh$.

9. Jeff builds a wooden cubic toy box for his son. Each side of the box measures 2 feet. How many square feet of wood does he use to build the toy box? How many cubic feet of toys will the box hold?

12.17   Front, Top, Side, and Corner Views of Solid Objects

Solid objects are 3-dimensional and therefore, are able to be viewed from several perspectives. You should be able to recognize the corner view of a solid given the front, top, and side views. Likewise, you should be able to draw and/or recognize the front, top, and side views of an object given its corner view.

Example 11:   Draw the front, top, and side view of the object shown below.

Solution:

Top   Front   Side

Example 12:   The top, front, and side views of an object are shown below. How many cubes would it take to build this object?

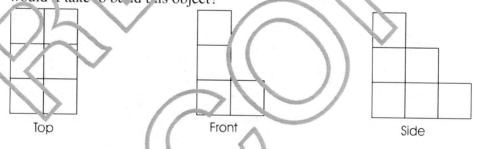

Top   Front   Side

Solution: Draw the object first, and then count the number of cubes use to create the structure.

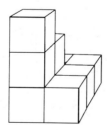

9 cubes are needed to build this object.

Copyright © American Book Company

Refer to the object shown below to answer questions 1, 2, and 3.

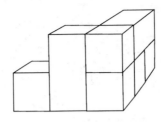

1. Which of the following is the top view of the solid?

(A) 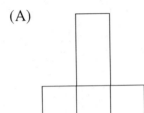 (B)  (C)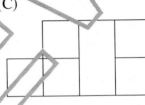

2. Which of the following is the side view of the solid?

(A) 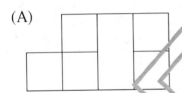 (B)  (C)

3. Which of the following is the front view of the solid?

(A)  (B)  (C)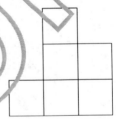

4. Given below are the front, top, and side views of a solid. Draw the object.

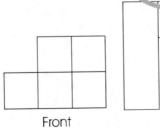 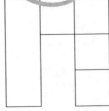 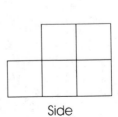

Front   Top   Side

Refer to the front, top, and side vies of the object shown below to answer questions 5 and 6.

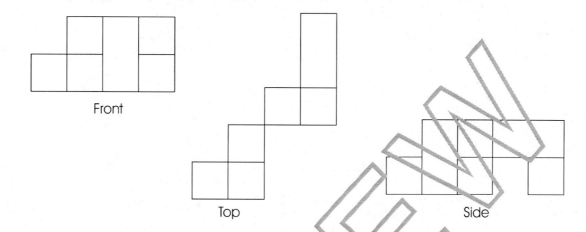

5. How many rectangular boxes would be needed to build this object?

6. How many cubes would be needed to build this object?

7. Given the object shown below, draw the front, top, and side views on your own paper.

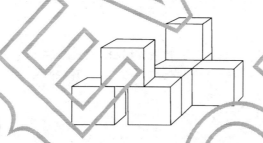

8. The front, side, and top views of a solid are show below. How many total blocks are needed to construct this object?

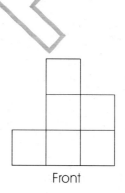

Front

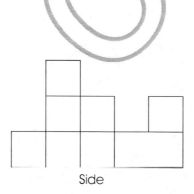

Side

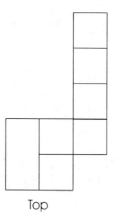

Top

12.18   Compare and Contrast Prisms and Pyramids

Recall the volume and surface area of rectangular prisms and pyramids.

Rectangular Prism

Pyramid

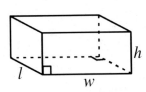

$$V = lwh$$
$$SA = 2hl + 2hw + 2lw$$

$$V = \tfrac{1}{3}lwh$$
$$SA = hl + hw + lw$$

The volume of a rectangular prism is three times the volume of a pyramid.
The surface area of a rectangular prism is twice as large as the surface area of a pyramid.
You can also write these relationships in another way:
The volume of a pyramid is one-third the size of the volume of a rectangular prism.
The surface area of a pyramid is half as large as the surface area of a rectangular prism.

Example 13:   The volume of a rectangular prism is 45 cm³. What the volume of a pyramid with the same dimensions?

Step 1:   Looking at the formulas above, we see that the volume of a rectangular prism is three times the volume of a pyramid. Let x be the volume of a pyramid, since we do not know the volume yet, and we will set up an equation. $3x = 45$ cm³

Step 2:   Solve the equation.

$$\frac{3x}{3} = \frac{45}{3} \text{ cm}^3 \qquad \text{Divide both sides by 3 to get } x \text{ by itself.}$$

$$x = 15 \text{ cm}^3 \qquad \text{The volume of a cone with the same dimensions is 15 cm}^3.$$

Answer the questions below by using the relationship between prisms and pyramids.

1. A rectangular prism has a volume of 30 ft³. What is the volume of a pyramid with the same dimensions?

2. A pyramid has a surface area of 90 m². What is the surface area of a rectangular prism with the same dimensions?

3. A rectangular prism has a surface area of 24 in². What is the surface area of a pyramid with the same dimensions?

4. A pyramid has a volume of 7 cm³. What is the volume of a rectangular prism with the same dimensions?

5. A rectangular prism has a width of 2 ft, a length of 8 ft, and a height of 3 ft. What is the volume of a pyramid with the same dimensions?

6. A pyramid has a width of 6 ft, a length of 6 ft, and a volume of 48 ft³. What is the height of a prism with the same width and length and has volume of 144 ft³?

12.19   Compare and Contrast Cylinders and Cones

Recall the volume and surface area of cylinders and cones.

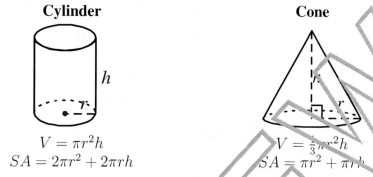

Cylinder

$$V = \pi r^2 h$$
$$SA = 2\pi r^2 + 2\pi r h$$

Cone

$$V = \tfrac{1}{3}\pi r^2 h$$
$$SA = \pi r^2 + \pi r h$$

The volume of a cylinder is three times the volume of a cone.

The surface area of a cylinder is twice as large as the surface area of a cone.

You can also write these relationships in another way:

The volume of a cone is one-third the size of the volume of a cylinder.

The surface area of a cone is half as large as the surface area of a cylinder.

Example 14:   The surface area of a cone is 16 in². What is the surface area of a cylinder with the same dimensions.

Step 1:   Looking at the formulas above, we see that the surface area of a cone is half as large as the surface area of a cylinder. Let x be the surface area of the cylinder, since it is the unknown, and make a formula to solve for x. $\tfrac{1}{2}x = 16$ in²

Step 2:   Solve the equation.
$2\left(\tfrac{1}{2}x\right) = 2\left(16 \text{ in}^2\right)$ Multiply both sides of the equation by 2 to get x by itself.
$x = 32$ in²   The surface area of a cylinder with the same dimensions is 32 in²

Answer the questions below by using the relationship between cylinders and cones.

1. A cone has a volume of 6 in³. What is the volume of a cylinder with the same dimensions?

2. A cylinder has a surface area of 42 in². What is the surface area of a cone with the same dimensions?

3. A cone has a surface area of 56 cm². What is the surface area of a cylinder with the same dimensions?

4. A cylinder has a volume of 21 m³. What is the volume of a cone with the same dimensions?

5. A cone has a radius of 4 ft, and a height of 3 ft. What is the volume of a cylinder with the same dimensions?

6. A cylinder has a diameter of 2 ft, and a volume of 9 ft³. What is the height of a cone with the same diameter and has volume of 3 ft³?

Chapter 12 Review

Find the volume and/or surface area of the following solids.

1.

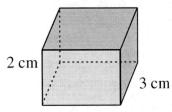

2 cm

3 cm

3 cm

$V =$ _____
$SA =$ _____

2.

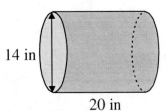

14 in

20 in

Use $\pi = \frac{22}{7}$.
$V =$ _____
$SA =$ _____

3.

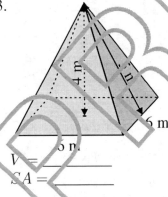

4 m

6 m

$V =$ _____
$SA =$ _____

4.

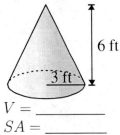

6 ft

3 ft

$V =$ _____
$SA =$ _____

5.

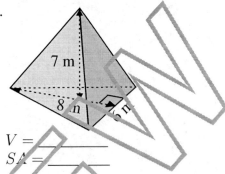

7 m

8 m

$V =$ _____
$SA =$ _____

6.

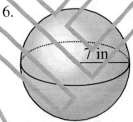

7 in

Use $\pi = \frac{22}{7}$.
$V =$ _____
$SA =$ _____

7. The sandbox at the local elementary school is 60 inches wide and 100 inches long. The sand in the box is 6 inches deep. How many cubic inches of sand are in the sandbox?

8. If you have cubes that are two inches on each edge, how many would fit in a cube that was 16 inches on each edge?

9. If you double each edge of a cube, how many times larger is the volume?

10.

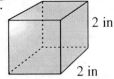

2 in

2 in

2 in

It takes 8 cubic inches of water to fill the cube below. If each side of the cube is doubled, how much water is needed to fill the new cube?

11. If a ball is 4 inches in diameter, what is its surface area? Use $\pi = 3.14$.

12. A grain silo is in the shape of a cylinder. If the silo has an inside diameter of 10 feet and a height of 35 feet, what is the maximum volume inside the silo? Use $\pi = \frac{22}{7}$.

13. A closed cardboard box is 30 centimeters long, 10 centimeters wide, and 20 centimeters high. What is the total surface area of the box?

14. Siena wants to build a wooden toy box with a lid. The dimensions of the toy box are 3 feet long, 4 feet wide, and 2 feet tall. How many square feet of wood will she need to construct the box?

15. How many 1-inch cubes will fit inside a larger 1 foot cube? (Figures are not drawn to scale.)

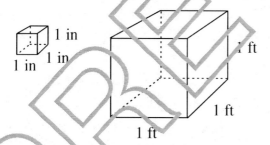

16. The cylinder below has a volume of 240 cubic inches. The cone below has the same radius and the same height as the cylinder. What is the volume of the cone?

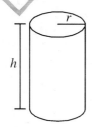 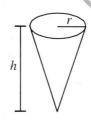

17. Estimate the volume of the figure below.

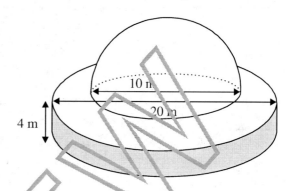

18. Find the volume of the figure below.

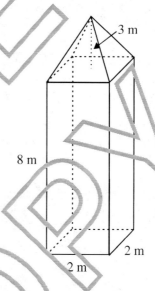

19. Find the volume of the figure below. Each side of each cube measures 4 feet.

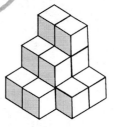

20. A gigantic bronze sphere is being added to the top of a tall building in Downtown. The sphere will be 24 ft in diameter. What will be the surface area of the globe?

Chapter 13
Symmetry

This chapter covers the following Georgia Performance Standards:

M6G	Geometry	M6G1.a, b
M6P	Process Skills	M6P1.b
		M6P3.d
		M6P4.c
		M6P5.b

Many geometric figures are symmetrical or have **symmetry**. Geometric figures can have three types of symmetry: **reflectional**, **rotational**, and **translational**.

13.1   Reflectional Symmetry

A figure has reflectional symmetry if you can draw a line through the figure that divides it into two mirror images. The mirror image line is called the line of symmetry. Look at the figures below.

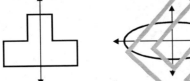

1 line of symmetry   2 lines of symmetry   4 lines of symmetry   infinite number of lines of symmetry

13.2   Rotational Symmetry

A figure has **rotational symmetry** if the image will lie on top of itself when rotated through some angle other than 360°. Look at the figures below.

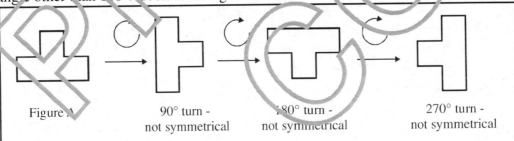

Figure A   90° turn -   180° turn -   270° turn -
                    not symmetrical   not symmetrical   not symmetrical

Figure A does not have rotational symmetry. It cannot be rotated so that the rotated image falls exactly on top of the original image.

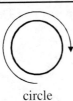

circle

A **circle** has complete rotational symmetry. No matter how much you rotate a circle, the rotated image will always look identical to the original circle.

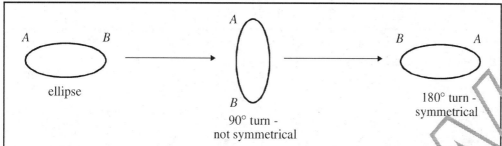

An **ellipse** has rotational symmetry at each   turn. An ellipse rotated   turn looks exactly like the original ellipse. All the points would lie exactly on top of each other.

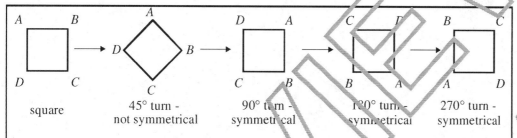

A square has rotational symmetry at each 90° turn. A square rotated 90°, 180°, or 270° looks identical to the original image.

13.3   Translational Symmetry

A geometric pattern has **translational symmetry** is an image can be slid a fixed distance in opposite directions to obtain the same pattern.

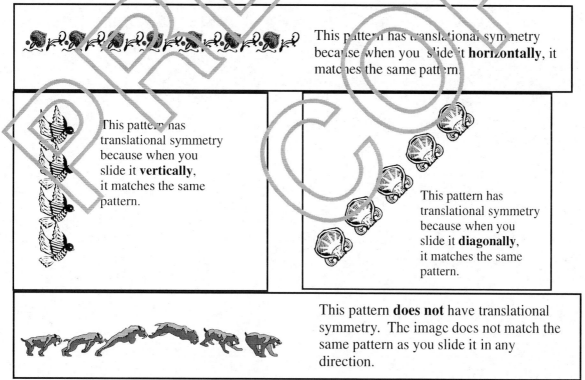

This pattern has translational symmetry because when you slide it **horizontally**, it matches the same pattern.

This pattern has translational symmetry because when you slide it **vertically**, it matches the same pattern.

This pattern has translational symmetry because when you slide it **diagonally**, it matches the same pattern.

This pattern **does not** have translational symmetry. The image does not match the same pattern as you slide it in any direction.

13.4   Symmetry Practice

Match each figure below to the letter that describes its symmetry. Some have more than one answer. Choose all the letters that apply.

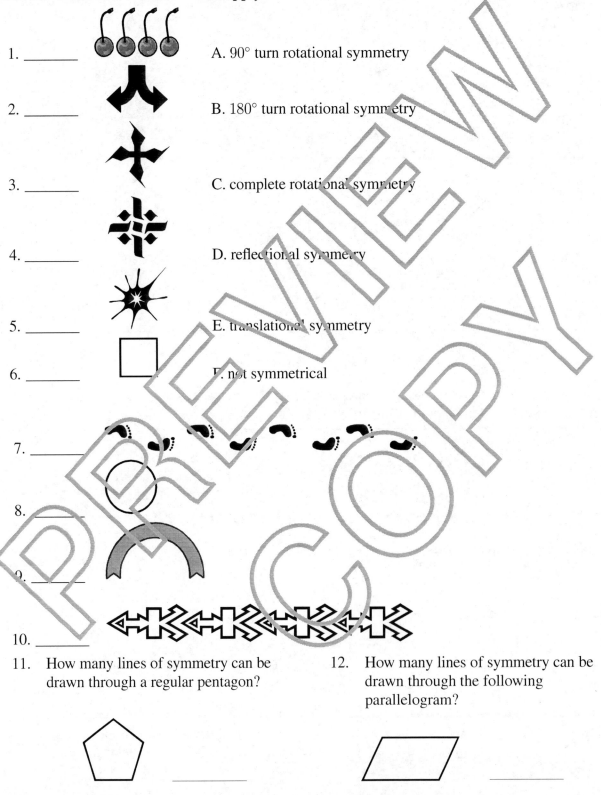

1. _____

A. 90° turn rotational symmetry

2. _____

B. 180° turn rotational symmetry

3. _____

C. complete rotational symmetry

4. _____

D. reflectional symmetry

5. _____

E. translational symmetry

6. _____

F. not symmetrical

7. _____

8. _____

9. _____

10. _____

11. How many lines of symmetry can be drawn through a regular pentagon?

12. How many lines of symmetry can be drawn through the following parallelogram?

Chapter 13 Review

Record the kind(s) of symmetry, if any, shown in each example below.

1.

2.

3.

4.

5. What kind(s) of symmetry does the following figure have? Choose the best answer.

    (A) reflectional symmetry
    (B) rotational symmetry
    (C) reflectional and rotational symmetry
    (D) no symmetry

6. What kind(s) of symmetry does the following figure have? Choose the best answer.

    (A) 90° rotational symmetry
    (B) 180° rotational symmetry
    (C) complete rotational symmetry
    (D) no symmetry

Formula Sheet

Perimeter	Rectangle	$P = 2l + 2w$ or $P = 2(l + w)$
Circumference	Circle	$C = 2\pi r$ or $C = \pi d$
Area	Rectangle	$A = lw$ or $A = bh$
	Triangle	$A = \frac{1}{2}bh$ or $A = \frac{bh}{2}$
	Trapezoid	$A = \frac{1}{2}(b_1 + b_2)h$ or $A = \frac{(b_1 + b_2)h}{2}$
	Circle	$A = \pi r^2$
Surface Area	Cube	$S = 6s^2$
	Cylinder (lateral)	$S = 2\pi rh$
	Cylinder (total)	$S = 2\pi rh + 2\pi r^2$ or $S = 2\pi r(h + r)$
	Cone (lateral)	$S = \pi rl$
	Cone (total)	$S = \pi rl + \pi r^2$ or $S = \pi r(l + r)$
	Sphere	$S = 4\pi r^2$
Volume	Prism or Cylinder	$V = Bh*$
	Pyramid or Cone	$V = \frac{1}{3}Bh*$
	Sphere	$V = \frac{4}{3}\pi r^3$
	*B represents the area of the base of a solid figure	
Pi	π	$\pi \approx 3.14$ or $\pi \approx \frac{22}{7}$
Pythagorean Theorem		$a^2 + b^2 = c^2$
Distance Formula		$d = \sqrt{(x_2 - x_1)^2 + (y_2 - y_1)^2}$
Slope of a Line		$m = \frac{y_2 - y_1}{x_2 - x_1}$
Midpoint Formula		$M = (\frac{x_2 + x_1}{2}, \frac{y_2 + y_1}{2})$
Quadratic Formula		$x = \frac{-b \pm \sqrt{b^2 - 4ac}}{2a}$
Slope-Intercept Form of an Equation		$y = mx + b$
Point-Slope Form of an Equation		$y - y_1 = m(x - x_1)$
Standard Form of an Equation		$Ax - By = C$

Practice Test 1

1. Sunshine Feed Supply has a huge cylindrical container that holds approximately 4,000 cubic feet of corn. The diameter is 10 feet. What is the approximate height of the container?

   (A) 13 feet
   (B) 40 feet
   (C) 50 feet
   (D) 100 feet

   M6M3c

2. The circular cone below has a base with a diameter of 8 cm and height 10 cm.

   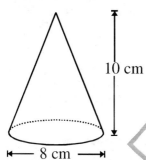

   The volume of the cone is most nearly

   (A) 42 cm³
   (B) 168 cm²
   (C) 503 cm³
   (D) 670 cm³

   M6M3b

3. The two trapezoids below are similar.

   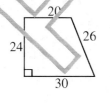 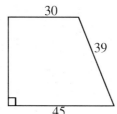

   What is the height of the larger trapezoid?

   (A) 16
   (B) 30
   (C) 32
   (D) 36

   M6G1c

4. Sally makes a scale drawing of her garage.

   The actual length of the garage is 21 feet. What is the width of the garage?

   (A) 10 feet
   (B) 11 feet
   (C) 12 feet
   (D) 14 feet

   M6G1e

5. A preschool is required to have a playground of at least 900 square feet. Which of the following would be satisfactory measurements for a playground for the school?

   (A) 30 feet by 32 feet
   (B) 27 feet by 30 feet
   (C) 15 feet by 40 feet
   (D) 10 feet by 80 feet

   M6M2c

6. Of the 410 visitors at the museum on Saturday, 164 are students. What percent of the visitors are NOT students?

   (A) 30
   (B) 40
   (C) 50
   (D) 60

   M6N1g

7. Which of the following can be used to compute $\frac{5}{6} + \frac{2}{9}$?

   (A) $\dfrac{5}{6 \times 9} + \dfrac{2}{6 \times 9}$

   (B) $\dfrac{5 + 2}{6 + 9}$

   (C) $\dfrac{5 \times 3}{6 \times 3} + \dfrac{2 \times 2}{9 \times 2}$

   (D) $\dfrac{5}{6 \times 3} + \dfrac{2}{9 \times 2}$

   M6N1d

8. A box contains spools of thread: 3 spools of red, 4 spools of blue, 2 spools of green, and 3 spools of yellow. What is the probability of reaching in the box without looking and picking a red spool?

(A) $\frac{1}{4}$

(B) $\frac{1}{5}$

(C) $\frac{2}{7}$

(D) $\frac{3}{4}$ M6D2b

9. Which of the following is an accurate graph of the point values below?

Length	Perimeter
1	3
2	6

(A)
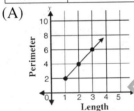

(B)
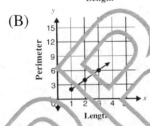

(C)
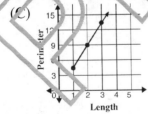

(D)
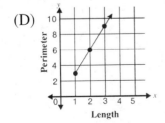

M6A2a

10. A Ferris wheel has a radius of 14 feet. How far will you travel if you take a ride that goes around six times? Use $\pi = \frac{22}{7}$.

(A) 528 feet
(B) 616 feet
(C) 3,696 feet
(D) 12,936 feet   M6M2b

11. To find the height of a flag pole, Emily holds a yardstick at right angles to the ground. The yardstick (36 inches) cast a 12-inch shadow, while the flag pole cast a 17-foot shadow. What is the approximate height of the flag pole?

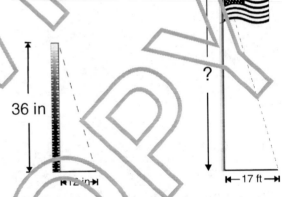

36 in

NOTE: Figures are not drawn to scale.

(A) 50 ft
(B) 55 ft
(C) 60 ft
(D) 65 ft   M6G1c

12. Joseph goes to a fast food restaurant and orders a hamburger for $1.79, fries for $1.39, and a large shake for $1.99. What is a reasonable amount to give to the cashier?

(A) $5.00
(B) $11.00
(C) $3.00
(D) $6.00   M6N1g

13. 14.2 is the same as

(A) $\frac{142}{100}$

(B) $14\frac{1}{50}$

(C) $14\frac{1}{5}$

(D) $14\frac{1}{10}$

M6N1f

14. $6\frac{1}{4} - 3\frac{7}{8} =$

(A) $2\frac{5}{8}$

(B) $2\frac{1}{2}$

(C) $2\frac{3}{8}$

(D) $3\frac{5}{8}$

M6N1d

15. 5 miles per hour is the same as how many feet per second?

(A) $\frac{5\,(5280)}{1}$

(B) $\frac{5\,(5280)}{2\,(60)}$

(C) $\frac{5\,(5280)}{(60)}$

(D) $\frac{5\,(5280)}{(60)^2}$

M6M1

16. This year, $\frac{7}{8}$ of all the 6th graders have signed up to go to the Valentine's Day dance. What percent of the students will be going to the dance?

(A) 0.78%

(B) 0.875%

(C) 8.75%

(D) 87.5%

M6N1f

17. Jason wants to know how many honey bees he has in his hive. On Day 1, he lures 40 bees into a trap and colors their wings with a harmless green dye and releases them back into the hive. On Day 2, he lures 10 bees into the trap and finds 2 bees have dyed wings. About how many bees must be in the hive?

(A) 80

(B) 100

(C) 200

(D) 400

M6A2g

18. A bricklayer uses 7 bricks per square foot of surface to be covered. If he covers an area 28 feet wide and 10 feet high, about how many bricks will he use?

(A) 40

(B) 2,000

(C) 2,800

(D) 3,000

M6M4c

19. Which figure below represents a cylinder?

(A)

(B)

(C)

(D)

M6G2d

20. Use the tree diagram below to predict the probability of flipping one coin 3 times and getting one head and two tails.

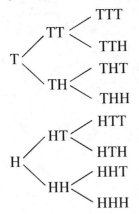

(A) $\frac{1}{2}$

(B) $\frac{1}{4}$

(C) $\frac{3}{8}$

(D) 3

M6D2a

21. In a basketball shooting contest, which of the following players has the lowest percentage of shots made?

(A) Greg makes 2 out of 7 shots.
(B) Erica makes 60% of his shots.
(C) Bob makes $\frac{3}{8}$ of his shots.
(D) Kent makes 5 out of 8 shots.   M6N1f

22. Mary owns a cat named Snoopy. She reaches into her bag of 4 fish, 6 liver, 3 chicken-flavored, and 10 milk treats and gives one to Snoopy without looking. What is the probability that Snoopy gets a liver treat?

(A) $\dfrac{1}{6}$

(B) $\dfrac{6}{17}$

(C) $\dfrac{6}{23}$

(D) $\dfrac{1}{23}$ M6D2b

23. To make a disinfecting solution, Alana mixes 2 cups of bleach with 5 cups of water. What is the ratio of bleach to the total amount of disinfecting solution?

(A) 2 to 3
(B) 2 to 5
(C) 2 to 7
(D) 2 to 10   M6A1

24. Which of these is the best estimate of the volume of this box?

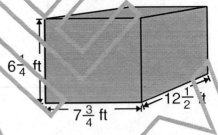

(A) 25 cubic feet
(B) 150 cubic feet
(C) 600 cubic feet
(D) 1200 cubic feet   M6M3c

25. Kenneth is planning to build a sailboat that will be 6.5 meters long. In the plans, the length of the sailboat is 130 mm. What is the ratio of the length of the actual sailboat to the length of the sailboat in the plans?

(A) 5 to 1
(B) 20 to 1
(C) 50 to 1
(D) 200 to 1   M6A1

26. A 10-pound bag of fertilizer will cover 50 square feet of garden space. How many 10-pound bags of fertilizer would you need to cover a garden that is 480 square feet?

(A) 9
(B) 10
(C) 20
(D) 48   M6A2c

27. It took Melanie 161 minutes to do her homework last night. About how many hours did she spend on her homework?

(A) $2\frac{2}{3}$ hours

(B) 2 hours

(C) $3\frac{1}{2}$ hours

(D) $1\frac{1}{2}$ hours   M6M1

28. What is an appropriate measure to use to show the distance from New York City to Miami, FL?

(A) kilometers

(B) meters

(C) grams

(D) liters   M6M1

29. Which of the following can be used to compute $\dfrac{1}{8} + \dfrac{3}{4}$?

(A) $\dfrac{1+3}{8+4}$

(B) $\dfrac{1}{8 \times 4} + \dfrac{3}{4 \times 4}$

(C) $\dfrac{1}{8 \times 1} + \dfrac{3}{4 \times 2}$

(D) $\dfrac{1 \times 1}{8 \times 1} + \dfrac{3 \times 2}{4 \times 2}$ M6N1d

30. Using the following table and assuming that the population at the art show continues to increase during the first hour, how many people will be at the art show one hour after the start of the show?

TIME IN MINUTES	NUMBER OF PEOPLE AT THE ART SHOW
0	12
10	22
20	32
30	42

(A) 52

(B) 62

(C) 72

(D) None of the above   M6A2a

31. Which of the following is equal to 30 meters?

(A) 0.3 millimeters

(B) 0.003 kilometers

(C) 300 centimeters

(D) 30,000 millimeters   M6M1

32. Darin is playing a dart game at the country fair. At the booth, there is a spinning board completely filled with different colors of balloons. There are 6 green, 4 burgundy, 5 pink, 3 silver, and 8 white balloons. Darin aims at the board with his dart and pops one balloon. What is the probability that the balloon popped is **not** green?

(A) $\frac{1}{13}$

(B) $\frac{10}{13}$

(C) $\frac{3}{13}$

(D) $\frac{2}{13}$ M6D2b

33. If you wanted to show the percentages of your allowance you spent in different categories, which graph would you choose to best represent the data?

(A) bar graph

(B) pictograph

(C) line graph

(D) circle graph   M6D1c

34. Which of the following is the prime factorization of 68?

(A) $2^2 \times 17$

(B) 2×34

(C) 4×17

(D) $12 \times 5 + 8$ M6N1b

35. In the triangle below, $\triangle ACE$ is similar to $\triangle BCD$. What is the measure of $\overline{AE}$?

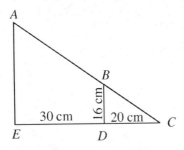

(A) 70 cm
(B) 60 cm
(C) 40 cm
(D) 24 cm

36. On a map drawn to scale, 2 centimeters represent 300 kilometers. How long would a line measure between two cities that are 500 kilometers apart?

(A) $1\frac{1}{5}$ centimeters
(B) $3\frac{1}{3}$ centimeters
(C) 5 centimeters
(D) $7\frac{3}{5}$ centimeters

37. Find the volume of the figure below. Each edge of each cube measures 4 feet.

(A) 56 ft³
(B) 226 ft³
(C) 896 ft³
(D) 3584 ft³

38. The starting balance for Robert was $210.00. He then wrote checks for $62.00, $35.80, $26.70, and $23.65. Later that week, he deposited $53.75. Then he withdrew $122.00. How much money does Robert have left in his account?

(A) −$6.40
(B) $6.40
(C) −$60.18
(D) $12.80

39. Dupree Park is adding a new playground area that measures 108 feet long and 48 feet wide. They want to add 3 inches of sand to the entire area. How many cubic yards of sand do they need to order?

(A) 12.5 cubic yards
(B) 48 cubic yards
(C) 144 cubic yards
(D) 1800 cubic yards

40. Below is shown a solid object constructed with cubes. Which of the following diagrams represents the side view of this object?

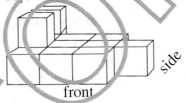

(A)
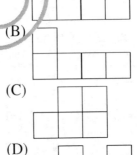

(B)

(C)

(D)
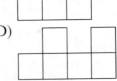

41. A cylinder with radius 3 inches and height 10 inches is filled with liquid. The liquid is poured into a cylinder with radius 5 inches. What is the height of the liquid in the second cylinder?

(A) 1.5 inches
(B) 2.8 inches
(C) 3.6 inches
(D) 6 inches

M6M3d

42. Bob is buying frozen fruit juice bars for $1.00 each and ice cream bars for $1.20 each for a picnic. He has $40.00 to spend. Which of the following graphs models the possible number of fruit juice bars and the possible number of ice cream bars Bob can buy?

(A)
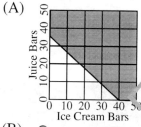

(B)
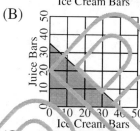

(C)
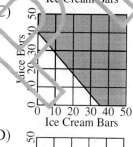

(D)
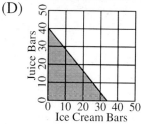

M6D1c

43. If this shape's sides were folded upward at the dotted lines, what three - dimensional object would it make?

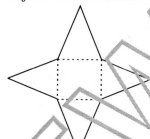

(A) cube
(B) cylinder
(C) cone
(D) pyramid

M6G2d

44. If 60 students eat 24 pizzas, which proportion below may be used to find the number of pizzas required to feed 15 students?

(A) $\dfrac{60}{24} = \dfrac{15}{x}$

(B) $\dfrac{60}{24} = \dfrac{x}{15}$

(C) $\dfrac{60}{15} = \dfrac{x}{24}$

(D) $\dfrac{60}{x} = \dfrac{15}{24}$

M6A2b

45. For every 3 hours Randall works, he takes a ten minute break. Randall worked 42 hours this week. How many minutes of break did he take?

(A) 140 minutes
(B) 14 minutes
(C) 420 minutes
(D) 200 minutes

M6A2c

46. Paul constructed a bar graph showing the number of each of four types of vehicle sold by Sunshine Motors last year.

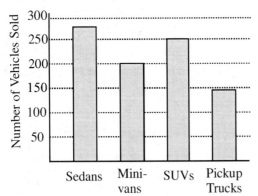

Approximately how many more sedans and minivans were sold than SUVs and pickup trucks?

(A) 25
(B) 50
(C) 75
(D) 100

M6D1d

47. A sweater is marked 25% off the retail price of $29.99. Robyn, who works at the store, will receive an additional 20% employee's discount off the sale price. How much will Robyn need to pay for his sweater?

(A) $13.50
(B) $15.00
(C) $16.49
(D) $17.99

M6N1g

48. 0.875 written as a fraction is

(A) $\frac{87}{100}$

(B) $\frac{4}{5}$

(C) $\frac{7}{8}$

(D) $\frac{22}{25}$

M6N1f

49. Find y: $-4y = 56$

(A) $y = -14$
(B) $y = 14$
(C) $y = -224$
(D) $y = 224$

M6A3

50. What is the area of the following circle? Use $A = \pi r^2$ and $\pi = 3.14$.

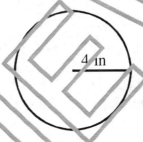

(A) 12.56 in²
(B) 16 in²
(C) 25.12 in²
(D) 50.24 in²

M6M2b

51. The area of the shaded region of the rectangle is 6 square feet. What is the length of the rectangle?

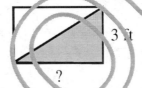

(A) 2 ft
(B) 3 ft
(C) 4 ft
(D) 6 ft

M6M2b

52. What would replace n in this number sentence to make the sentence true?
$\frac{1}{2}n = 16$

(A) 4
(B) 8
(C) 32
(D) 64

M6A3

53. The diagram below shows a box without a top. Each of the sides is a rectangle. What is the surface area of the five sides?

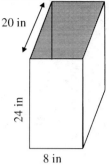

20 in

24 in

8 in

(A) 832 in^2
(B) 992 in^2
(C) 1504 in^2
(D) 1664 in^2

M6M4d

54. Below are shown the top, front, and side views of a solid. How many cubes are needed to build this 3-dimensional object?

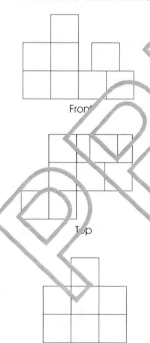

Front

Top

Side

(A) 9
(B) 10
(C) 11
(D) 12

M6G2c

55. What kinds of symmetry does the following figure have? Choose the best answer.

(A) 180° rotational symmetry
(B) reflectional symmetry
(C) translational symmetry
(D) no symmetry

M6G1b

56. How many lines of symmetry can be drawn through the following figure? Choose the best answer.

(A) 2
(B) 4
(C) infinite
(D) none

M6G1a

57. A cylinder and a cone have the same dimensions for their radius and heights. How does the volume of the cylinder compare to the volume of the cone?

(A) The volumes are the same.
(B) The volume of the cylinder is one-third the volume of the cone.
(C) The volume of the cylinder is three times as large as the cone.
(D) The volume of the cylinder is four times as large as the cone.

M6G2b

58. Lee works part-time at a fast-food restaurant and makes $45 per week. He saves 75% towards college. How much will he have saved after 32 weeks?

(A) $1,080.00
(B) $33.75
(C) $53.33
(D) Not enough information is given.

M6N1g

59. Hanna buys 3 pairs of socks priced at 3 for $5.00 and shoes for $45.95. She pays $2.55 sales tax. How much change does she receive from $100.00?

(A) $36.50
(B) $46.50
(C) $51.50
(D) $53.50

M6N1g

60. Multiply: $2\frac{3}{4} \times 1\frac{1}{5}$

(A) $2\frac{3}{20}$
(B) $2\frac{7}{24}$
(C) $3\frac{3}{10}$
(D) $6\frac{3}{5}$

M6N1e

61. What is the greatest common factor of 36 and 120?

(A) 2
(B) 6
(C) 12
(D) 18

M6N1c

62. What is the least common multiple of 6 and 8?

(A) 2
(B) 16
(C) 24
(D) 48

M6N1c

63. Kevin needs $3\frac{1}{3}$ yards of streamers to decorate each table. How many tables will 10 yards of streamers decorate?

(A) 3
(B) $3\frac{1}{3}$
(C) 10
(D) $66\frac{2}{3}$

M6N1e

64. What kind(s) of symmetry does the following figure have? Choose the best answer.

(A) 90° rotational symmetry
(B) 180° rotational symmetry
(C) complete rotational symmetry
(D) no symmetry

M6G1b

65. A pine tree casts a shadow 9 feet long. At the same time, a rod measuring 4 feet casts a shadow 1.5 feet long. How tall is the pine tree?

(A) 3.375 feet
(B) 13.5 feet
(C) 24 feet
(D) 54 feet

M6G1e

66. What are the first 3 multiples of 12?

(A) 1, 2, 3
(B) 2, 3, 4
(C) 1, 12, 24
(D) 12, 24, 36

M6N1a

67. What is the prime factorization of 32?

(A) 2^5
(B) 2×4^2
(C) $2^2 \times 8$
(D) 2×16

M6N1b

68. Which year did the sheep outnumber the goats by 1000?

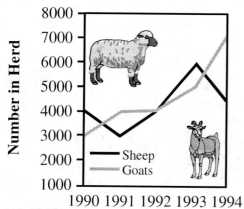

(A) 1991
(B) 1992
(C) 1993
(D) 1994

M6D1e

69. Frankie flips two coins at the same time for a total of 100 times each. He writes down what combination of heads or tails he gets each time. Below are his results.

HT	21	TH	29
HH	33	TT	17

According to Frankie's results, what is the probability for tossing a combination of TH? Express as a percent.

(A) 21%
(B) 29%
(C) 33%
(D) 62%

M6D2a

70. A neighborhood surveyed the times of day people water their lawns and tallied the data below.

Time	Tally
midnight - 3:59 a.m.	II
4:00 a.m. - 7:59 a.m.	IIII I
8:00 a.m. - 11:59 a.m.	IIII IIII
noon - 3:59 p.m.	IIII
4:00 p.m. - 7:59 p.m.	IIII IIII
8:00 p.m. - 11:59 p.m.	IIII III

What time of the day was the most popular time to water the lawn?

(A) midnight - 3:59 a.m.
(B) 8:00 a.m. - 11:59 a.m.
(C) 4:00 p.m. - 7:59 p.m.
(D) Not enough information is given

M6D1b

Practice Test 2

1. The price of a skateboard increased from $32.80 to $39.00. What is the approximate percentage of increase?

   (A) 19%
   (B) 16%
   (C) 8%
   (D) 6%   M6N1g

2. The bar graph below shows the number of members of the cheerleading team at four grade levels.

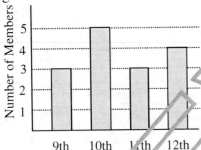

   Which of the following statements is not a correct interpretation of the bar graph?

   (A) 30% of members of the cheerleading team are in the 9th grade.
   (B) There are 5 members of the cheerleading team in the 10th grade.
   (C) There are 15 members on the cheerleading team.
   (D) The 9th and 11th grades have an equal number of members.   M6D1e

3. Which is the equivalent multiplication problem for $\frac{3}{4} \div \frac{2}{3}$?

   (A) $\frac{4}{3} \times \frac{3}{2}$

   (B) $\frac{3}{4} \times \frac{2}{3}$

   (C) $\frac{4}{3} \times \frac{2}{3}$

   (D) $\frac{3}{4} \times \frac{3}{2}$ M6N1e

4. Patty has carefully weighed and measured the length of a licorice stick before taking the first bite and again after each bite. From her data shown in the table below, she has concluded that the weight of the remaining licorice stick is proportional to the length.

Bite Number	Length	Weight
0	304mm	28.6 grams
1	285mm	26.3 grams
2	259mm	23.5 grams
3	239mm	22.5 grams
4	202mm	? grams

   After the 4th bite, the licorice stick was 202 mm long. Approximately how many grams should the licorice stick have weighed?

   (A) 17 grams
   (B) 18 grams
   (C) 19 grams
   (D) 20 grams   M6A2a

5. Janice is comparing the price of three brands of olive oil. Which brand is the best buy?

Olive Oil	Size (milliliters)	Price
Brand X	709 mL	$10.99
Brand Y	500 mL	$8.65
Brand Z	442 mL	$4.99

   (A) Brand X is the least expensive per mL.
   (B) Brandy Y is the least expensive per mL.
   (C) Brand Z is the least expensive per mL.
   (D) Cannot be determined.   M6D1d

6. How many square feet is a 9-foot by 60-foot lawn?

   (A) 69 square feet
   (B) 138 square feet
   (C) 270 square feet
   (D) 540 square feet   M6M2b

7. What is the volume of a wading pool 12 feet long, 6 feet wide, and 6 inches deep?

(A) 18 cubic feet

(B) 36 cubic feet

(C) 216 cubic feet

(D) 432 cubic feet M6M3d

8. What is the area of a circle with a radius of 7 cm? (Round to the nearest whole number)

(A) 154 square cm

(B) 196 square cm

(C) 347 square cm

(D) 616 square cm M6M2a

9. Bonnie makes toy wagons to sell to gift shops. She buys wheels for the wagons at \$12.72 per dozen. Which formula will calculate the cost of 4 wheels per wagon?

(A) $12x = \$12.72$

(B) $\dfrac{12}{4}x = 12.72$

(C) $\dfrac{12.72}{12} = x$

(D) $4x = 12.72$ M6A2d

10. $\frac{1}{2} - \frac{1}{3} =$

(A) $\frac{1}{6}$

(B) -1

(C) 0

(D) $\frac{1}{5}$ M6N1d

11. Jenna needs one foot square floor tiles for her bathroom. Her bathroom is 6 feet by 5 feet. The floor tiles come 11 tiles to a box. About how many boxes does she need?

(A) 2

(B) 3

(C) 4

(D) 5 M6M4d

12. If 3 out of 4 people take a specific vitamin, how many in a city of 150,400 will take this vitamin?

(A) 118,200

(B) 37,600

(C) 50,133

(D) 112,800 M6A2f

13. The scale of a map is $\frac{1}{2}$ inch = 40 miles. If two towns are 6 inches apart on the map, how many miles apart are they?

(A) 480 miles

(B) 120 miles

(C) 20 miles

(D) 30 miles M6G1e

14. What is the measure of side x?

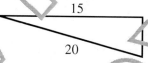

(A) 8

(B) 9

(C) 10

(D) 16 M6G1c

15. $\frac{4}{5} + \frac{1}{6} =$

(A) $\frac{5}{11}$

(B) $\frac{5}{30}$

(C) $\frac{29}{30}$

(D) $\frac{29}{11}$ M6N1d

16. On a recent test, 80% of the math class got A's. What fraction of the class is that?

(A) $\frac{1}{4}$

(B) $\frac{5}{6}$

(C) $\frac{5}{8}$

(D) $\frac{4}{5}$ M6N1f

17. If it takes Nathan and his friends 6.5 hours to travel 300 miles to Memphis for a concert, which of the following proportions could help him figure out how long it would take to go 650 miles to the group's next concert?

(A) $\dfrac{6.5}{T} = \dfrac{650}{300}$

(B) $\dfrac{650}{6.5} = \dfrac{300}{T}$

(C) $\dfrac{6.5}{300} = \dfrac{T}{650}$

(D) $\dfrac{300}{650} = \dfrac{T}{6.5}$

M6A2b

18. $4\frac{5}{6}$ written as a decimal is

(A) 0.456
(B) 4.56
(C) $4.8\overline{3}$
(D) $0.48\overline{3}$

M6N1f

19. Diego bought a box of nails of various sizes. In the box were 30 three-inch nails, 40 two-inch nails, 60-one inch nails, and 20 half-inch nails. If he dumps out the nails on the table and one falls off, what is the probability that a one-inch nail falls off the table?

(A) $\frac{1}{2}$

(B) $\frac{1}{60}$

(C) $\frac{2}{5}$

(D) $\frac{3}{5}$

M6D2b

20. Alphonso saw a stereo on sale for $\frac{1}{3}$ off the regular price of $630. How much money could he save if he bought the stereo on sale?

(A) $210
(B) $410
(C) $600
(D) $630

M6N1g

21. Victoria recorded how far her pet snail crawled. After 10 minutes the snail had crawled 20 centimeters. After 15 minutes, it had crawled 30 centimeters. Assuming the snail crawls at a constant rate, which of these graphs shows the distance traveled by the snail as a function of time?

(A)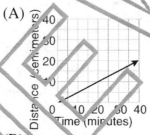

(B)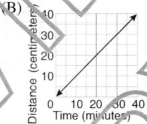

(C)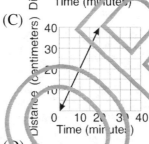

(D)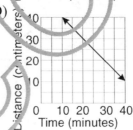

M6D1e

22. Which of the following is the prime factorization of 90?

(A) $2 \times 3^2 \times 5$
(B) 30×3
(C) 15×6
(D) $2 \times 3 \times 15$

M6N1b

23. What is the volume, in cubic feet, of the square pyramid below?

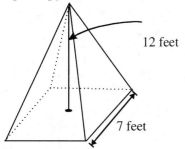

12 feet

7 feet

(A) 168 cubic feet
(B) 196 cubic feet
(C) 294 cubic feet
(D) 588 cubic feet

M6M3b

24. A box of candy contains 3 chocolate mint, 5 chocolate nut, 4 taffy, 6 butterscotch, and 2 vanilla candies. If one piece of candy is selected at random, what is the probability that it will contain chocolate?

(A) 0.15
(B) 0.25
(C) 0.4
(D) 0.5

M6D2b

25. Madison is reading the floor plans of her new house. What is the perimeter of the room shown below?

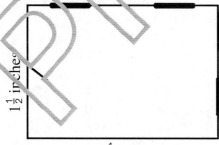

$1\frac{1}{2}$ inches

$2\frac{1}{4}$ inches

Scale: $\frac{1}{8}$ inch $= 1$ foot

(A) 20 feet
(B) 60 feet
(C) 40 feet
(D) 30 feet

M6G1d

26. A backyard is 160 feet wide. What is the width of the field in yards?

(A) 50 yards
(B) 53 yards
(C) $53\frac{1}{3}$ yards
(D) $53\frac{2}{3}$ yards

M6M1

27. $\frac{3}{4} \times \frac{3}{4} \times \frac{3}{4} =$

(A) $\frac{27}{64}$

(B) $\frac{3}{4}$

(C) $\frac{6}{7}$

(D) $\frac{27}{4}$

M6N1e

28. If all corresponding sides and all corresponding angles of two triangles are congruent, then they are congruent triangles. If two triangles are congruent, then they are similar triangles. Given these facts, which of the following statements is valid?

(A) Similar triangles have all sides and all angles congruent.
(B) If two triangles are similar, then they are congruent.
(C) If two triangles are not congruent, then they are not similar.
(D) If two triangles have all corresponding sides and angles congruent, then they are similar triangles.

M6G1c

29. Which of the following choices is the best estimate for the width of a standard door?

(A) 9 meters
(B) 9 kilometers
(C) 90 centimeters
(D) 90 millimeters

M6M1

30. What is the volume of the figure below?

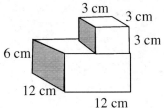

3 cm 3 cm
3 cm
6 cm
12 cm
12 cm

(A) 21 cm
(B) 39 cm
(C) 216 cm
(D) 891 cm

M6M3d

31. Which of the following figures represents a pyramid?

(A)
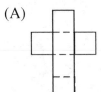

(B)
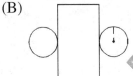

(C)

(D)

M6C2d

32. There are 20 male and 35 female students taking band this year at Washington Middle School. What is the ratio of female students to male students taking band this year?

(A) $\frac{4}{7}$

(B) $\frac{7}{4}$

(C) $\frac{7}{11}$

(D) $\frac{11}{7}$

M6A1

33. A box of candy measures 6 inches by 3 inches by 2 inches. What is the volume of the box in cubic inches?

(A) 22 cubic inches
(B) 11 cubic inches
(C) 42 cubic inches
(D) 36 cubic inches

M6M3d

34. Sabrina wants to cover a box on all six sides with white satin. The box is 4"×6"×10". If she glues the fabric on so it does not overlap, how many square inches of fabric will she use?

(A) 120 square inches
(B) 124 square inches
(C) 240 square inches
(D) 248 square inches

M6M4d

35. Find the missing side from the following similar triangles.

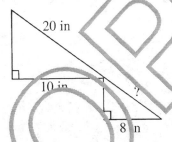

20 in

10 in
?
8 in

(A) 8
(B) 16
(C) 24
(D) 32

M6G1c

36. Dawn makes $7.50 per hour. She worked 36 hours last week. Twenty percent of her income was taken for taxes. Of the remaining amount, she saves $\frac{1}{3}$. How much money does she have left to spend?

(A) $72
(B) $180
(C) $36
(D) $144

M6N1g

37. On a scale drawing $\frac{1}{2}$ inch represents 1 foot. A line segment is 11 inches long. How many feet are represented by the line segment?

(A) 11 feet
(B) 22 feet
(C) 44 feet
(D) 55 feet

M6G1d

38. Solve: $-\frac{4}{5}x = 8$

(A) $x = -2$
(B) $x = 5$
(C) $x = -10$
(D) $x = 40$

M6A3

39. Johnny is playing a game in which he picks up one rubber duck out of a pool of floating rubber ducks. Each duck has a prize printed on the bottom of the duck. Four ducks show a stuffed animal prize, 7 show a plastic jewelry prize, 5 show a squirt gun prize, and 11 show a spinning top prize. What is the probability that Johnny will select a duck with a stuffed animal prize?

(A) 11 out of 27
(B) 4 out of 27
(C) 7 out of 11
(D) 1 out of 27

M6D2b

40. Out of 480 students, $\frac{3}{5}$ bought hot lunches on Monday. How many students bought hot lunches?

(A) 96
(B) 160
(C) 288
(D) 800

M6N1g

41. Which of the following sets contains equivalent numbers?

(A) $\frac{9}{25}$ 0.35   35%
(B) $\frac{5}{16}$ 0.315 $31\frac{1}{2}\%$
(C) $\frac{3}{8}$ 0.375 $37\frac{1}{2}\%$
(D) $\frac{4}{5}$ 0.08   80%

M6N1f

42. The cheerleaders buy 50 boxes of doughnuts to sell after school. They pay $62.50 for all the boxes, then sell them for $2.00 per box. How much profit do they make on each box?

(A) $0.62
(B) $0.75
(C) $1.00
(D) $1.06

M6N1g

43. Brad's shop class is making a doll house as a group project. Brad sketches the part he is about to cut.

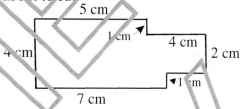

The teacher says it is too small and tells Brad to redraw the figure with the 4 cm sides lengthened to 7 cm each. If he redraws the figure and makes sure it is similar, how long with the 5 cm line be?

(A) 6 cm
(B) 8 cm
(C) 8.75 cm
(D) 9.25 cm

M6G2e

44. Gerald works in a garment factory. He can sew on 3 buttons in 13 seconds. At this rate, approximately how many buttons would Gerald sew on in one hour?

(A) 130
(B) 270
(C) 800
(D) 1300

M6A2c

45. What is the prime factorization of 128?

(A) $2^3 \times 42$
(B) 2^7
(C) $2^2 \times 32$
(D) $2^6 \times 3$

M6N1b

46. Janice is in a job training program. For each 5 hours she works, 3 hours are paid by her employer and 2 hours are paid by the training program. For last week's work, the job training program paid Janice for 12 hours. How many hours did Janice work last week?

(A) 18 hours
(B) 24 hours
(C) 30 hours
(D) 40 hours

M6A2c

47. Sasha has a variety of plants and animals in her fish tank. She has 5 multicolored delta guppies, 4 silver angelfish, 6 clown fish and 1 dalmation molly. What is the ratio of silver angelfish to the total number of fish?

(A) 1 : 1
(B) 1 : 3
(C) 4 : 13
(D) 1 : 4

M6A1

48. According to the phone bill, Pam spent 224 minutes talking to her boyfriend, John, last month. About how many hours was that?

(A) $2\frac{1}{3}$ hours

(B) $2\frac{3}{4}$ hours

(C) $3\frac{3}{4}$ hours

(D) 22 hours

M6M1

49. The area of a picture is 192 in². What is the area of the picture in square feet?

(A) 12 ft²
(B) $\frac{3}{4}$ ft²
(C) $1\frac{1}{3}$ ft²
(D) 16 ft²

M6M1

50.

Favorite Lunch Item	Frequency
corndog	140
hamburger	245
hotdog	210
pizza	255
spaghetti	90
other	65

Based on the data in the chart, a student chosen at random is most likely to want which two choices for lunch?

(A) a corndog or pizza
(B) a hotdog or spaghetti
(C) a hamburger or pizza
(D) a hotdog or pizza

M6D1e

51. What is the greatest common factor of 72 and 54?

(A) 2
(B) 3
(C) 9
(D) 18

M6N1c

52. What is the LCM of 2 and 10?

(A) 2
(B) 5
(C) 10
(D) 20

M6N1c

53. Sheila has some carpet that is $8\frac{1}{2}$ feet long. She uses 5 pieces that are $\frac{3}{4}$ feet long. What is the length of the carpet remaining?

(A) 5 feet

(B) $4\frac{3}{4}$ feet

(C) $3\frac{3}{4}$ feet

(D) $7\frac{3}{4}$ feet

M6N1d

160

54. Find the value of x. $\dfrac{x}{1.5} = 6$

   (A) 4
   (B) 6
   (C) 9
   (D) 90   M6A3

55. Which of the following is a multiple of 6?

   (A) 2
   (B) 3
   (C) 16
   (D) 24   M6N1a

56. Which of the following is a factor of 32?

   (A) 3
   (B) 8
   (C) 9
   (D) 64   M6N1a

57. Which is the prime factorization of 28?

   (A) $2^2 \times 7$
   (B) 2×14
   (C) 4×7
   (D) $2^2 \times 14$ M6N1b

58. What kind of symmetry does a square have?

   (A) line symmetry
   (B) 180° rotational symmetry
   (C) 90° rotational symmetry
   (D) All of the above   M6G1a

59. Which of the following formulas would you use to find the volume of a pyramid?

   (A) $V = lwh$
   (B) $V = \frac{1}{3}lwh$
   (C) $V = \pi r^2 h$
   (D) $V = \frac{1}{3}\pi r^2 h$ M6M3a

60. What is the surface area of a cylinder that has a height of 4 cm and a diameter of 2 cm? (Round to the nearest whole number.)

   (A) 30 cm^2
   (B) 31 cm^2
   (C) 32 cm^2
   (D) 75 cm^2 M6M4b

61. How many blocks would be needed to create the object given the top, front, and side views?

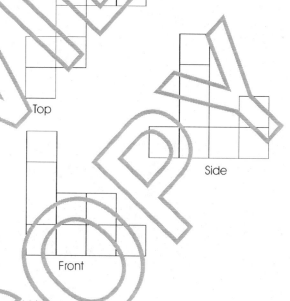

   (A) 9
   (B) 10
   (C) 11
   (D) 12   M6G2c

62. The volume of a cone is 150 cm^3. A cylinder has the same diameter and height as the cone. What is the volume of the cylinder?

   (A) 50 cm^3
   (B) 150 cm^3
   (C) 450 cm^3
   (D) Cannot be determined   M6G2b

63. Which is the graph of the equation $y = 3x$?

(A)

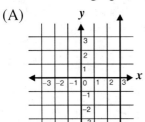

(B)

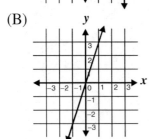

(C)

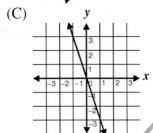

(D)

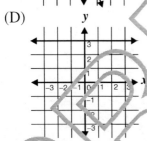

M6A2e

64. What type of graph would you use to show the change of shoe size over time?

(A) line graph
(B) bar graph
(C) circle graph
(D) pictograph   M6D1c

65. What is the probability of getting a heads when flipping a coin?

(A) 2

(B) $\frac{1}{4}$

(C) $\frac{1}{2}$

(D) 1   M6D2b

66. Sally rolled a die 20 times and got a four 3 times. Based on this information, what is the experimental probability of rolling a four?

(A) $\frac{1}{6}$

(B) $\frac{3}{20}$

(C) $\frac{6}{20}$

(D) Cannot be determined   M6D2a

67. The length of a rectangle is 5, and the length of another rectangle is 12. If these two rectangles are similar, what is the scale factor to get from the first rectangle to the second rectangle?

(A) 1.4
(B) 2.4
(C) 7
(D) $\frac{5}{12}$ M6G1c

68. Find the value of x. $\frac{x}{-4} = 2$

(A) $\frac{1}{2}$
(B) $-\frac{1}{2}$
(C) 8
(D) -8 M6A3

69. The volume of a rectangular prism is 45 in³, and the volume of a similar pyramid is 15 in³. If the width and the height of the rectangle is 3 and 5 inches, what is the length of the pyramid?

(A) 1 in
(B) 3 in
(C) 5 in
(D) Cannot be determined   M6G2a

70. What are the first 3 factors of 32?

(A) 1, 2, 4
(B) 2, 4, 6
(C) 32, 64, 96
(D) 64, 96, 128   M6N1a

Index

Acknowledgements, ii
Algebra
  one-step problems with addition and subtraction, 66
  one-step problems with multiplication and division, 67
Area Problems, Two-Step, 107
Arguments, 62
  inductive and deductive, 63

Bar Graphs, 81
Borrowing in Subtraction, 24

Cartesian Plane, 71
Circle
  area
    $A = \pi r^2$, 107
Circle Graphs, 83
Conclusion, 62
Congruent Figures, 110
Contrapositive, 64
Converse, 64
Corresponding Sides
  of similar triangles, 111
Counterexample, 64
Customary Measure, 98

Decimal Word Problems, 36
Decimals
  adding, 30
  changing to fractions, 35
  changing to percents, 38
  changing with whole numbers to mixed numbers, 36
  division by decimals, 33
  division by whole numbers, 32
  multiplication, 32
  subtracting, 31
Deductive Reasoning, 62
Denominator, 21

Diagnostic Test, i
Diameter, 106
Discount, Finding the Amount of, 46
Discounted Sale Price, Finding, 47

Evaluation Chart, 13

Formula Chart, ix, 142
Fractions
  adding, 22
  changing to decimals, 34
  changing to percents, 39
  comparing the relative magnitude of, 27
  dividing, 26
  improper, 18
  multiplying, 25
  reducing, 20
  word problems, 26
Frequency Table, 78
Front, Top, Side, and Corner Views of Solid Objects, 130

Geometric Relationship of Solids, 120
Graphs
  bar, 81
  circle, 83
  line, 82
  linear data, 75, 86
  pictographs, 84
  pie, 83
Greatest Common Factor, 16

Histogram, 79

Improper Fractions
  simplifying, 18
Inductive reasoning, *see* Patterns, 62
Inverse, 64

Key (Legend), 84

Least Common Multiple, 17

Legend (Key), 84
Line Graphs, 82
Linear Equation, 71
Logic, 62

Maps, 54
Mass, 99
Mathematical Reasoning, 62
Metric System, 99
  converting units, 101
Mixed Numbers
  changing to decimals, 35
  changing to improper fractions, 19
  changing to percents, 40
  subtracting from whole numbers, 23
  subtracting with borrowing, 24

Negative Numbers
  multiplying and dividing with, 68
Nets
  of solid objects, 127
  using to find surface area, 128
Numerator
  finding missing, 21

One-Step Algebra Problems
  addition and subtraction, 66
  multiplication and division, 67

Patterns
  inductive reasoning, 58
  number, 57
Percent of Increase and Decrease, Finding, 44
Percent of Total, Finding, 43
Percent Word Problems, 42
Percents
  changing to decimals, 38
  changing to fractions, 39
  changing to mixed numbers, 40
Perimeter, 103
$\pi = 3.14$ or $\pi = \dfrac{22}{7}$, 106
Pictographs, 84
Polygon, 103
Population, 87

Practice Test 1, 143
Practice Test 2, 154
Preface, viii
Premises
  of an argument, 62
Prime Factorization, 14
Probability, 90
Proportional Reasoning, 53
Proportions, 51
Proposition, 62

Radius, 106
Ratio and Proportion Word Problems, 52
Ratios, 50, 51
Rectangle
  area
    $A = lw$, 104
Reducing Fractions, 26
Relative Magnitude of Numbers, 41

Sales Tax, 48
Sample, 87
Scale Drawings, 54
Similar Figures, 110
Simulation, 93
Slope
  $m = \dfrac{y_2 - y_1}{x_2 - x_1}$, 73
Square
  area
    $A = lw$, 104
Surface Area
  cone, 126
  cube, 122
  cylinder, 125
  pyramid, 124
  rectangular prism, 122
  solids, 122
  sphere, 126
Surveys, 87
Symmetry
  reflectional, 138
  rotational, 138
  translational, 139

Table of Contents, vii
Tables
   data, 80
Tally Chart, 78
Tips and Commissions, 45
Triangles
   area
$$A = \frac{1}{2} \times b \times h, 105$$
   similar, 111
Two-Step Area Problems, 107

Variable, 66
   coefficient of negative one, 70
Volume
   cube, 116
   liquid, 99
   rectangular prism, 115

solids, 114
spheres, cones, cylinders, and pyramids, 117
two-step problems, 119

Whole Numbers
   subtracting mixed numbers, 23
Word Problems
   decimals, 36
   fractions, 26
   percent, 42
   ratios and proportions, 52
   solid geometry, 130

x-axis, 76

y-axis, 76

165